STEAM POWER IN AGRICULTURE
IN COLOUR

STEAM POWER
IN AGRICULTURE

Michael Williams

drawings by
Denis Bishop

BLANDFORD PRESS
POOLE   DORSET

First published 1977
by Blandford Press Ltd,
Link House, West Street, Poole, Dorset BH15 1LL

Copyright © Blandford Press Ltd 1977

To Sylvia

ISBN 0 7137 0819 0

Printed in Great Britain by
Fletcher & Son Ltd, Norwich

CONTENTS

ACKNOWLEDGEMENTS

To those who have helped in any way with the preparation of this book, please accept my thanks. In particular I would like to mention the following: staff at the Museum of English Rural Life, University of Reading, for their unfailing helpfulness; Allan Herman of Portland, Oregon, and Frank Hollis of International Harvester Company of Great Britain for their help in providing original material for pictures. Unless otherwise indicated the photographs were supplied by Peter Adams.

M.W.

STATIONARY ENGINES

Steam in Britain

The first attempts to put steam power to work in agriculture marked the beginning of one of the greatest revolutions in farming. Since that time, at the end of the eighteenth century, the tireless efficiency of engine power has been gradually replacing the drudgery of human and animal effort.

The revolution is still far from complete. In large areas of the world, farming is still geared to the pace of the peasant and his oxen, just as it was two centuries ago in Britain. Of the progress so far achieved, most has been due to the internal-combustion engine. Steam pioneered the principles of power farming, but the internal-combustion engine has taken over to make the power revolution a practical reality.

Steam power was already firmly established in Britain for operating pumps in mines, and for some other industrial uses, before serious attempts were made to link steam to farming. Thomas Newcomen, a blacksmith from Dartmouth, Devonshire, designed the first practical steam-engine, and Newcomen engines were working in several mining areas by the time their inventor died in 1729. This type of engine could have had little value in farming. It used great quantities of both coal and water in relation to the modest power output, and the engines were costly to buy and to install. Another disadvantage was that the Newcomen engine was designed to deliver its power only through a reciprocating motion – suitable for operating a pump – and not with a rotary motion which might power gears or a pulley.

James Watt and others improved on Newcomen's design during the eighteenth century, raising the efficiency and power output, and paving the way for steam power to be applied to boats and railways in the nineteenth century.

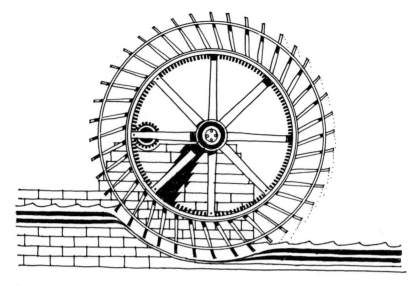

Sectional drawing of the Martin Mere paddle wheel of 1850. (See p. 18.)

Meanwhile there were inventors and theorists attempting to devise ways of using steam for farming. Mostly they planned to use steam in the field, especially for ploughing and cultivating. This was somewhat impractical in the eighteenth century, when steam power was firmly stationary, and portable or self-propelled engines were still years away.

Credit for being first to suggest farming by steam power, is sometimes given to David Ramsey and Thomas Wildgosse who jointly patented an idea for ploughing without the use of horses or oxen. Their patent refers vaguely to 'Newe, apte, or compendious formes or kindes of engines or instruments and other pfitable invenciones'. It is not at all clear exactly what Messrs. Ramsey and Wildgosse had in mind, but there is little reason to believe that it was steam. Their patent was filed in 1618, almost a century before Newcomen had made steam a practical source of power. In 1618 steam was just an interesting curiosity, and the Ramsey–Wildgosse idea was more probably based on the energy in gunpowder, or perhaps on the hope of some suitable power source being developed eventually. There was, however, an indication in a later patent taken out by Ramsey, that fire was involved in powering the invention.

During the next 150 years there were other suggestions and patents aimed at replacing horses and oxen for field work, but these were not obviously based on steam, and none appeared to advance beyond theory.

Francis Moore took out a series of patents in 1769, all referring to his invention of a machine to replace horses and oxen. His device was to be powered by fire, water or air, and would need the assistance of two men. The inventor claimed that the machine would plough the soil, and would also do a wide range of cultivating operations. There is some evidence that Francis Moore actually built a machine based on his patents, and that it worked. Presumably the prototype failed in some way to impress, and nothing more was heard of it. But it did impress its inventor. He was so sure of success that he sold his own farm horses, confident that the popularity of his new machine would make so many draft horses redundant that their price would fall. Generously, Mr Moore encouraged his neighbours to follow his example while their horses still had value.

The optimistic Mr Moore might have found it easier to attempt to use steam to operate stationary equipment on the farm. The reason why he, and others before him, concentrated on field equipment was that there was little in the way of barn machinery for the steam-engine to drive. The machinery that was available on most farms demanded little power, and was either hand operated or powered by oxen or horses working a system of gears.

This situation changed in the 1780s when the threshing drum was developed, principally by Andrew Meikle in Scotland. British farming then, and for the next thirty years or so, enjoyed high prosperity. There was widespread interest and enthusiasm for new ideas and improvements, together with the capital to invest in them. The threshing drum was welcomed as a replacement for the laborious method of threshing with flails, and its use spread surprisingly quickly to most areas of the country.

It was the threshing drum which brought steam power to farming. Farmers demanded cleaner threshing and higher outputs, and drum sizes were increased to produce the required results. The bigger drums needed larger teams, and it was difficult to achieve the constant, steady power which gave best results.

The first attempts to replace animal power by using a steam-engine took place around the turn of the eighteenth century. Fortunately for the historian it was at this time that British farming came under the scrutiny of a detailed, county-by-county survey, which continued during the early years of the nineteenth century. Through these surveys it is possible to trace with reasonable accuracy the slow introduction and spread of the steam-engine into different areas of the country. This is in spite of the fact that some of those responsible for the surveys were not interested in farm implements and mechanization.

According to the surveys the first farm in Britain, and almost certainly the first in the world, to use steam power was in North Wales, in what was formerly the county of Denbighshire. Mr W. Davies, the author of *A General View of the Agriculture of North Wales*, gave little information about the dawn of power farming. He simply recorded that the owner of the steam-engine was named Wilkinson, and that he had first used it successfully to power his threshing drum in 1798. In a later edition of the survey, Wilkinson is described as an ironmaster, and in fact it is almost certain that he was John Wilkinson, a businessman with a factory at Wrexham and a farm nearby. John Wilkinson was a business associate of James Watt, who was then manufacturing steam-engines in partnership with Matthew Boulton at the famous Soho works in Birmingham. Wilkinson held the patent for a process for accurately machining metal cylinders. This was used mainly for making better cannons and guns, but was also used to make some precision parts for Boulton and Watt engines. Wilkinson bought several engines from the Soho works for his own factory, including one of the first with Watt's sun and planet gears to produce a rotary motion. This engine, installed in the 1780s, could have been removed later to the farm.

A year later, in 1799, a steam-engine was used to power a threshing drum on a farm in East Lothian, Scotland. The owner was referred to elsewhere as 'a respectable farmer'. The author of *A General View of the Agriculture of the County of East Lothian* gives us little detail of the installation, apart from mentioning that as the engine was so close to a coal-mine, the cost of fuel was low. Elsewhere in the report, the cost of coal at the pithead is quoted as 4½*d.* (approximately 2p) per hundredweight.

While the use of threshing machines was spreading quickly among the larger farms and estates, the idea of using steam power to drive them was accepted more cautiously. In 1803 Arthur Young, one of the most respected of farming writers and the author of several of the county surveys, reported only one engine on a farm in Norfolk, and none in the county of Suffolk. The Norfolk engine, which was being installed in 1803 by Col. Buller on his farm at Heydon, was described by Young as probably the first in Britain to be used specifically for farm work. Col. Buller was planning to use the engine to power his threshing drum, a grinding mill and a chaff cutter – simultaneously if necessary. This would replace ten horses. The engine was expected to grind nine bushels of wheat, while running the threshing machine and the chaff cutter, at a cost of 84 lb of 'good Newcastle coals'. The complete installation, including the threshing and grinding equipment, was expected to cost £600. This was at a time when a skilled Norfolk ploughman might earn £12 a year plus his food and bed. In *A General View of the Agriculture of the County of Norfolk* Young gave only one paragraph to Col. Buller's steam-engine. This suggests that steam on the farm was of little interest to him, as he wrote at great length and in fine detail about anything he considered important.

In *A General View of the Agriculture of the County of Durham*, the author, J. Bailey, proved to be one of the few writers of his day to express enthusiasm for the idea of farming with steam power. On the subject of threshing he wrote: 'The expense of horses is now become so great, that they should not be used where it can be avoided. Water, where it can be obtained, is certainly the cheapest. Wind, tho' an uncertain power, is in many instances used; but in this county, where coals are so cheap, and in most cases at no great distance, steam is probably the best and most effective power that can be employed, where the farm is sufficiently large to require a powerful machine.'

There are two steam-powered threshing machines described in the Durham survey. One, at Pallion, had been installed to pump water from a limestone quarry, and the owner also used it for threshing. The other engine had been installed in 1805 on a farm at Chillingham Barns, over the border in Northumberland. The Chillingham engine was rated at 5 h.p., but the owner claimed it was more than equal to six of his horses. The engine worked a large threshing machine with a 6 ft drum, and output was claimed to be 1,000 to 1,200 sheaves threshed per hour. Coal consumption was claimed to be twenty to twenty-four bushels for six hours' work, with coal in the area costing 10s. (50p) for twenty-four bushels. The cost of the installation was £100 for the threshing machine, and £325 for the steam-engine including the building in which it was operated. Writing five years after the engine had been installed at Chillingham, Mr Bailey claimed that it had worked every year without problems, and with only minor maintenance costs.

Boulton and Watt steam-engines dominated the industry until the early years of the nineteenth century when James Watt's patents expired. Most of the farm engines so far referred to would have been of this design, which gave modest power output in relation to the size of the engine, its cost and the amount of fuel used. The Watt engine operated at little more than atmospheric pressure and made slight use of the expansive force of steam. The building referred to in the costings for the Chillingham engine was probably not a shelter for the operator, but a substantial structure of stone which would have been an essential part of the framework and support for the engine.

Watt's main patents expired in 1800, and one of the engineers who took advantage of the fact was Richard Trevithick, born 1771, the son of a Cornish mining engineer. Trevithick made a number of important contributions to steam-engine improvement, including the outstanding development of an engine which would work at relatively high pressures. An essential feature of this development was the use of boilers of cylindrical shape to operate at higher pressures than the rectangular box boilers of the Watt engines. Another feature of the 'Cornish' type of boiler produced by Trevithick was locating the fire-box actually inside the boiler. Under the genius of Trevithick, the steam-engine became much more compact, less expensive to install and to run, and more efficient than previous designs.

Richard Trevithick was interested in developing the use of steam power for farming, and his ideas included a steam-powered rotary cultivator. One of his stationary engines was installed in 1811 at Home Farm, Trewithin, Probus, Cornwall for the owner, Sir Christopher Hawkins.

The engine cost £80, and was used for threshing. It was kept on the farm until 1879, when a far-sighted descendant of the original owner donated the engine to the Science Museum in London. In that year the engine was included in a display of historical equipment which was part of the Royal Agricultural Society's show at Kilburn, London. The R.A.S.E. Journal for 1879 devoted several pages to a description of the engine, which was apparently one of the great attractions at the show. The engine, fitted with a Cornish boiler of slightly more modern design than the original would have been, is still in the possession of the museum, although at the time of writing it was dismantled and not displayed.

With more efficient engines available, British agriculture began to accept steam power more readily. More barn equipment, such as root and chaff cutters, bean and oil cake crushers were linked to steam-engines, and farming writers became more enthusiastic. There was much encouragement for steam to be used to 'cook' food for pigs and to steam potatoes and chaff for dairy cows. In 1861, H. Stephens and R. S. Burn, joint authors of *The Book of Farm Buildings* suggested that waste heat from steam-engines should be used for drying grain – either in the sheaf or after threshing. By 1880 this suggestion had become hard fact, and the farming press of the time carried descriptions and pictures of portable steam-engines providing the heat and the power to operate drying equipment for both hay and grain.

J. B. Denton, in his book *Farm Homesteads of England* written in 1863, described a 614-acre farm in Herefordshire in which the buildings had been planned around a centralized stationary steam-engine. The 12-h.p. engine provided the power for a comprehensive range of barn machinery for processing feed for the cattle, which were housed nearby. Metal rails had been laid in the yards and buildings and specially designed trolleys travelled on these tracks to bring sheaves from the stackyards to the threshing machine located beside the engine. The engine also provided the power to work an auger for carrying away the threshed straw.

This planned use of steam power must have surprised those who had earlier failed to see a future for steam on the farm. There were many who shared the view of Capt. Thomas Williamson, that steam power could not be economically used on most farms. In his book, *Agricultural Mechanism*, published in 1810, the Captain declared: 'I do not see how steam-engines can become generally serviceable to the farmer, unless, indeed, he be also a brewer or a maltster, in which case he may find means to heat his steam kettle, without any additional expence of fuel.'

Steam for American Farms

The threshing machine, which created the first real demands for steam power on British farms, also created a big market for steam-engines in North America. But this situation developed in American farming long after it had become firmly established in Britain. It was the plantation owners in the southern states who first brought steam power into American farming, using the new engines to drive their sugar-cane crushers, their cotton gins and rice mills, and to saw timber from their woodlands.

It is not possible to give details of where and when the first engines were used for farm work in America, but there is some evidence that there were stationary engines on plantations in the south by 1810, and there is also evidence that sugar-cane and cotton growers were quick to appreciate the advantages of steam. Water-wheels, windmills, tidemills and animal power all had limitations, and demand for steam-engines increased rapidly in the southern states, and also in the sugar-producing islands of the West Indies.

The development of the use of steam power in the United States has been researched and reported by Carroll W. Pursell in his book, *Early Stationary Steam Engines in America* (Smithsonian Institution Press, 1969). This book gives some details of negotiations in 1812 between a planter from Louisiana, and Benjamin Latrobe, who was one of the pioneers of American steam-engine production. Latrobe quoted $2,500 as a fair price for a small engine of 12 in. diameter cylinder, to power a sugar mill. On this figure he was not expecting to make a profit, but was hoping to pay for castings, patterns and tools. He anticipated further orders for similar small stationary engines for sugar estate work, and expected that these would develop into more profitable business.

During the first twenty years of the last century, the steam-engine business in America was very much in its infancy. There was a lack of experience, of skilled engineers and metalworkers, plus problems of transport and communications and war. In this situation some of the demand for steam power in America, and especially in the south, was met by British manufacturers. British engines were also supplying much of the steam power for estates in the islands of the Caribbean.

The American Secretary to the Treasury in 1838, Levi Wood-bury, published a comprehensive survey for that year, showing the numbers of stationary steam-engines, steamboats and steam loco-motives in America, on a state-by-state basis. This famous survey indicates how rapidly the number of stationary engines had in-creased in the south. Louisiana then had 274 stationary engines – more than any other state excepting Pennsylvania. The survey also indicated the total and the average horsepower figures for the engines in each state, and in many cases quoted the country of origin of the engines. There were Boulton and Watt engines from

England in several of the southern states, including seven in Louisiana and three in Georgia. Rice was an important crop in the state of South Carolina, and Carroll Pursell's book states that one of the first rice mills to be powered by steam was in Savannah in 1815, when a 90-h.p. Boulton and Watt engine from England was imported. The engine was later moved to an electricity generating plant in Savannah where it was still operating in 1894.

The popularity of steam power for operating sugar mills continued to increase after the 1838 survey, when about 140 of the mills in Louisiana were worked by steam. By 1844 it was estimated that more than 400 mills in the state were steam powered, with only 354 still worked by animal power. In 1861 over 1,000 steam-engines were operating Louisiana sugar mills.

Although British manufacturers continued to sell small numbers of engines to America, and especially to the British islands in the West Indies with sugar estates, American manufacturers were quick to make up for their slow start, both in volume and quality. In the 1850s it was reported that almost all of the steam-engines on Cuban sugar estates were American built, and twenty years later American stationary engines were exported to Britain, where some were advertised for farm use.

Land Drainage

One of the first uses for steam power was pumping water from mines and quarries, and by the 1850s some landowners in Britain were harnessing steam to pump water for land drainage. The earliest steam-powered land-drainage schemes were in Lincolnshire, but one of the biggest and most successful installations was an 1,100-acre project on Martin Mere, Southport, Lancashire, on part of the estate belonging to Sir Thomas Hesketh. The work was completed in 1850, and was described in the *Journal of the Royal Agricultural Society*, Vol. 14, in an article by Henry White, a land agent from Warrington.

The Martin Mere scheme was centred on a stationary steam-engine driving a vertical scoop wheel. The wheel, 30 ft in diameter, had 16-in. wide paddles around its circumference, and these forced water 'uphill' along a 16·5 in. wide channel. The wheel worked rather like a waterwheel in reverse, and turned at 4·25 r.p.m. – its most efficient speed. Water forced up to the outlet drain by the wheel was raised through a height of about 7 ft. It was estimated that the annual rainfall on the 1,100 acres amounted to 3,937,541 tons of water, of which 34 per cent was lost by evaporation and absorption, and 66 per cent was collected in the drainage system and lifted by the engine and paddle wheel. The engine was rated at 20 h.p., and was linked to a 30-h.p. boiler, 26 ft long, 6 ft diameter and with a 2 ft 3 in. diameter centre flue. The engine and scoop wheel, manufactured by Benjamin Hick and Son of Bolton, cost £1,025 complete with boiler. The total cost of the scheme including drains, channels, embankments and bridges was £3,425.

Running costs for the engine included 200 tons of coal at £51 delivered each year, and the 'engine man' cost £39 a year for wages, rent of cottage and two acres, and including the value of free coal for his home.

According to the R.A.S.E. *Journal*, the drainage scheme significantly improved the productivity of the marsh land. The improvement raised the value of 800 acres of the marsh from £529 before drainage, to £1,278 after. 'I cannot myself contemplate the improvement which has been here effected without feelings of the greatest satisfaction,' wrote the author.

'The poorer classes have here had productive labour provided for them, where, before, labour was comparatively fruitless. Land which was swampy, sterile, and unfit for human habitation, is now dry, productive and healthy. Works of this kind are not often undertaken; when they are, and successfully carried out, I think we confer a public good in making known the means by which such works have been affected.'

The Martin Mere scheme was described as the most successful of its type in Britain at that time, but it was on a tiny scale by comparison with a pumping scheme in Holland. This was to drain a big area of low-lying ground near Haarlem. Work started in 1843, and was described in several articles in the *Farmer's Magazine* published at that time in Britain. The scheme appears to have involved three stationary engines. The first of these was installed in 1845, and drove eleven pumps, each of 63 in. diameter. The engine and pumps were manufactured in England by Fox & Co. of Perran, Cornwall, and the total weight of engine, boiler and pumps was 640 tons. This installation was said to be able to lift one million tons of water through 10 feet in 25·5 hours.

With steam power transforming industry and transport, and beginning to make its impact on farming, there was growing excitement in the middle of the nineteenth century about the potential for further development of steam. One of the more unlikely schemes was described in 1856 by Mr W. Bridges Adams.

His theory, explained at some length in a paper read on his behalf at a meeting of the Society of Arts in London, involved the installation of pipes to take steam from a stationary or portable engine, below the surface of the soil. The benefits of doing this might, according to Mr Adams, be considerable. One result, he hoped, would be that the steam would raise the temperature of the soil to encourage plant growth during the winter. In his paper he anticipated the criticism that encouraging plants to grow in winter by warming their roots, might simply result in frost damage to the leaves. This might not happen, he thought, because the plants grown in this way might adjust to the new situation.

Warming the soil would do more than benefit plants, said Mr Adams. The underground heating below paths in gardens and parks would make walking in winter more enjoyable.

As the steam passed below ground, it could carry with it plant nutrients in gaseous form. These gases might simply be the products of combustion in the engine providing the steam, or they might be chemicals especially introduced into the steam, to be forced into the soil below the surface. In the opinion of Mr Adams, this would be of immense value in avoiding the great weight and transport cost of farmyard and other manures.

Just in case there were still those who doubted the value of piping steam underground, Mr Adams had one more theory to persuade them. Why not use the steam to cultivate the soil from below? 'It is just possible that the steam process may fissure the ground more advantageously from below; and thus the cost of digging and ploughing be saved in peculiar positions,' he explained.

The hopes and ambitions of those who tried to use the power of stationary engines for field work and farm transport were doomed. There are reports of temporary success, but these had little impact on commercial farming. The stationary engine established its place as the means to power stationary equipment. As engineering and production methods improved during the second half of the nineteenth century, scores of manufacturers in Europe and North America competed for a growing agricultural market. Engines and boilers became more efficient and economical, and more compact designs were introduced.

One of the first farmers to realize the scope of steam power was Alderman John J. Mecchi, of Tiptree, Essex. Mecchi overcame what, in nineteenth-century England was the very considerable handicap of a humble background, to make a fortune in industry. He invested some of his money in 170 acres of farmland, and became one of the leading innovators of his day, with a particular interest in farm mechanization. A paper given to the Farmers' Club in London in May 1855, included a description of Mecchi's use of steam for stationary work. He used an engine which was rated at 6 h.p. with 30 lb pressure, but operated at 70 lb to deliver 11 h.p. This engine was used to work a pair of 4 ft 4 in. millstones, a chaff-cutter, seed dresser, threshing machine and a linseed crusher. It powered a water pump, and also another pump with 100 gal. per minute output for spraying liquid manure from his cattle through an organic irrigation system. Finally, the engine also operated an air pump which was used to agitate and mix the liquid manure in the holding tank before the irrigation system was started.

Alderman Mecchi's farming was an outstanding demonstration of farming technology, and his example almost certainly encouraged progress in British agriculture. Mecchi also encouraged progress by using his own money to support what he considered to be promising developments. Unfortunately his grasp of agricultural science was not matched by his understanding of economics. It appears that his farming was geared to high costs with high output, and this was not the best system to face the agricultural depression of the 1870s. He ran into increasing financial problems, both in his farming and with his other sources of income, and he died in 1880 in circumstances as humble as those in which he had started.

Mecchi's example was followed, on a more modest scale, by many farmers who realized the value of the stationary engines which were being made available in rapidly increasing numbers. Some leading makers offered as many as twenty different sizes or models of engine, although the popular size for farm use was from 1 to 8 h.p. The leading agricultural shows, and the pages of the farming press of the time, were loud with the competing claims of manufacturers, often with lists of the medals and prizes won at shows and exhibitions around the world. This was the peak of popularity for the stationary steam-engine, for the shows and the advertisements were beginning to include the small petrol and paraffin engines which were soon to dominate the market.

2

STEAM ON WHEELS

As long as the steam-engine remained bolted firmly to the floor, its value on the farm was limited. A small number of farmers, such as Alderman Mecchi, tried with some success to justify the cost of a stationary engine on a farm with a modest acreage. Generally the economics were not attractive and stationary steam power was confined mainly to the big estates, until later in the nineteenth century when smaller engines with greater efficiency were available.

In order to make steam more versatile and more useful for farming, engines had to be movable. If an engine could be moved, it could be taken from farm to farm by a contractor, or else shared by a number of smaller acreage farms in a syndicate. The need for such a movable, or portable, engine was quite obvious, but engineers and farmers were surprisingly slow to get one designed. On both sides of the Atlantic steam was already proving its worth for propelling boats and powering railway locomotives. Some progress was also being made with self-propelled steam locomotives for road transport. Meanwhile farming made the most of the stationary engine, including some ingenious schemes for powering ploughs and other field equipment from a fixed engine.

One of the more desperate attempts to plough with the power of a stationary engine was described in a long article in the *Farmers' Magazine* of 1855. When the article was published there were already large numbers of portable engines in use, and some success was being achieved with cable ploughing systems, which meant that the idea of using a stationary engine for ploughing was already outdated.

The idea was described in relation to a farm of 640 acres, although presumably other farm sizes would have been suitable. The farm was rectangular, with the buildings at the centre and with the land fenced to form four fields, all rectangular and of equal size.

The central buildings would house livestock, feed preparation equipment and other barn machinery, and the steam-engine. From the buildings ran four sets of railway lines, laid along the boundaries between the four fields, and parallel with the railway there would be lines of shafting suitable for transmitting power. The railway was to be used to carry trucks, and these would be moved by power from the stationary engine transmitted through a cable system to pull the trucks in either direction.

Power from the centrally located engine would also be used to drive the shafting, and this would transmit power which would be applied to ploughing and to other field operations. The author of the scheme was either secretive or uncertain about the detail of how to power a plough from the pulleys on overhead shafting. 'The cultivation of the field, by means of shafting is a more problematical affair, yet easily done,' is all we are told.

The operation of the railway to give a complete transport system is more adequately described. The trucks would haul materials such as manure and seeds on the outward run, and they would be loaded with produce from the fields for the return journey to the buildings. The article suggested that some farmers would wish to connect their farm railway to the public system, presumably creating the opportunity for truckloads of potatoes or manure to be trundled on to the permanent way by unskilled hands.

Fortunately there were already portable and self-propelled steam-engines to provide farm transport and to cultivate the land. The first portable steam-engine for farm use was demonstrated publicly in 1841. A portable engine was little more than a stationary engine and boiler mounted on wheels, and with shafts at the front so that it could be pulled from place to place by horses, or very occasionally by oxen. This appears to have been quite a modest step forward in terms of technology, but its significance in making steam power more useful for agriculture was immense. In 1851, only ten years after the first portable was demonstrated, an estimated 8,000 had been sold to British farmers and contractors.

Although there is some doubt about who designed the first portable engine, the strongest evidence is that the credit should go to the Ipswich company, J. R. and A. Ransome, now known as Ransomes, Sims and Jefferies. The Ransomes' portable was on display at the Royal Agricultural Society's 1841 show, which was held at Liverpool that year, and it was also demonstrated on a farm at Fairford, near the showground, so that visitors to the show could see the engine in action.

The R.A.S.E. welcomed the Ransomes' exhibit, which probably attracted extra visitors to the show, and it was described in the Society's official report of the show as 'The great novelty of the meeting'.

The report continued: 'The advantages of steam-power for working fixed threshing mills have long been acknowledged in the northern parts of England and in Scotland; but we believe this is the first attempt to render it portable so that it may be transported from one farm to another.'

Ransomes mounted their engine on a two-wheel carriage; the complete unit weighed 35 cwt and was pulled by two horses. The engine was of the disc type, designed by a Mr Davies of Birmingham. At the Royal Show demonstration, the outfit was used to power a threshing drum. The performance was described as equivalent to five horses, in terms of power output, and fuel consumption was $1\frac{1}{2}$ cwt of coke an hour. Water consumption was measured at 36 gal. an hour, and a feature of the design was that the exhaust steam was piped through to the chimney where the condensing water was supposed to extinguish sparks and reduce the fire hazard.

Having introduced the portable engine in 1841, Ransomes made history again the following year when they turned their portable engine into what was probably the world's first self-propelled steam-engine for agriculture. This threshing engine was the forerunner of the agricultural traction engine.

It was the portable engine which had the greatest immediate impact. Within ten years there were more than a dozen manufacturers of portable engines in Britain, and it appears that some of these companies had already been working on the same idea when Ransomes announced their engine in Liverpool. In 1856, *The*

Engineer published an outline of the history of the portable steam-engine, and according to this report, the Ransomes engine had not been the first of its type. The article claimed that the first portable engine had been built by a Mr Dean of Birmingham, and he had completed the engine in 1841. This was the same year as the Ransomes' engine, but presumably the author believed that Mr Dean's engine had actually been completed before the Ransomes' engine.

The Engineer report also mentions a portable steam-engine which was completed by the Boston, Lincolnshire firm of Howden, also in 1841. The Howden engine was described as a 6-h.p. unit, with a single $8\frac{1}{2}$-in. diameter cylinder mounted vertically. The Howden engine was displayed at the Wrangle, Lincs, agricultural show.

The same source suggests that another Lincolnshire manufacturer, Tuxford of Boston, was probably the first to consider making a portable engine. The suggestion came from the manager of an experimental farm run at that time by the Earl of Ducie. The farm manager discussed his ideas with Tuxfords, and in 1839 he asked them to design and make a steam-engine which was to be combined with a threshing machine, the complete unit to be mounted on a wheeled chassis so that it could be moved easily from place to place. For some reason construction of the machine was delayed, and it was not completed until 1842.

From these reports it appears that there was considerable interest in Lincolnshire in the idea of using portable steam-engines for threshing. This is borne out by a reference published in the *Journal* of the R.A.S.E. for 1843, which indicated that contractors in the county were already using portable engines to operate a mobile threshing service for their customers. According to the report, there was already an agreed charge for contract threshing. This was between 1*s*. and 1*s*. 3*d*. (5p and 6p) per quarter of grain threshed, to pay for the engine plus the stoker and engineer. The farmer provided the remainder of the labour force at his expense, and he also provided coal for the engine. As well as agreeing a standard contract charge for the county, it appears from the R.A.S.E. that Lincolnshire contractors had also negotiated special insurance rates to cover the risk of fire when threshing with steam power.

Farmers in Britain welcomed the portable engine and appreciated its advantages, but the potential market for steam power was very much greater in North America. Farming in Britain was based predominantly on relatively small farms, with generally a balance between arable and livestock production, and adequate supplies of family or hired labour available. As farming spread westwards in the United States and Canada, with vast areas still to be settled, farm sizes were frequently very much greater than in Britain and Europe, and there were large areas of specialized production, especially of cereals. Labour and animal power were both in short supply to cope with cultivations and harvesting on the bigger holdings, and this situation created a big market for steam power by the end of the nineteenth century, and later an even bigger demand for tractors.

It appears that some portable steam-engines were produced in the United States on a one-off or experimental basis from about 1846, but details are not available, and it seems unlikely that any of these amounted to a commercial success. The first American to tackle the potential market seriously, was A. L. Archambault, from Philadelphia, Pennsylvania. His portable 'farm engine' was announced in 1849. It had a horizontal, locomotive-type boiler mounted on a four-wheel chassis, with the wooden wheels only 2 ft in diameter. There appears to be no record of the number of portables Archambault sold, but there is some evidence that he offered the 'farm engine' in three different sizes, with the choice of 4, 10 or 20 h.p.

Also in 1849 Charles Hoad and Gilbert Bradford of Watertown, New York state, began producing portable engines for farm use, on a small scale. A Hoad and Bradford engine was demonstrated at the 1851 New York State Fair, where the judges were sufficiently impressed to award it a medal and some words of praise on the convenience of its mobility.

These and other portable engines apparently failed to match all the requirements of farmers and contractors. In 1853 *The Scientific*

American claimed in a report that there was still nothing available to satisfy what they foresaw as the big potential demand for a portable engine which was both small and efficient. The report, quoted by Carroll Pursell in *Early Stationary Steam Engines in America*, suggested that there was a widespread need for such an engine, both for threshing and for manufacturing industry.

A large number of American manufacturers recognized the need for a portable engine, and during the years following the Civil War production increased rapidly. Some of the immense grain-producing farms in the west saw the portable steam-engine as their escape from the limitations of the horse sweep for powering threshing machines. The largest practical type of sweep or gear could take only fourteen horses, and the fast-expanding American harvest required much greater threshing capacity. On one farming property in the Red River Valley of Dakota, five portable engines were used for threshing spring wheat in 1877, and in 1884, when the wheat crop amounted to 30,000 acres, the number of portable engines reached thirty. On the Glenn Ranch in California in the early 1880s, there was a solid block of wheat stretching for sixteen miles and amounting to 66,000 acres. Several threshing records were established on this property, including 6,183 bushels of wheat threshed in one day between sunrise and sunset in 1879. The threshing machine was a 44-in. Gaar Scott, powered by a 25-h.p. Gaar Scott portable engine.

While the farmers with large acreages could justify the cost of a steam-powered threshing outfit, there were big areas of smaller properties where threshing offered opportunities for contract work. The contractors, or custom threshermen, were a substantial slice of the market for companies producing portable engines. This was particularly true in the eastern states, and there inadequate roads, steep hills and weak bridges often made the movement of the portable engine and threshing drum a difficult and time-consuming business. The portable engine was used to replace horses for threshing, but horses were still needed to move the engine, the water cart and the thresher. In spite of the difficulties, the custom threshing crew, with their portable engine soon to be replaced by a self-propelled unit, became a major factor in mechanizing the production of grain in the United States. The threshing outfit also became a notable feature in the popular history of rural America, and its annual appearance was both an economic and a social highspot in the year for thousands of farming families.

3

IN SEARCH OF SELF-PROPULSION

Farmers who were keen to make full use of steam power for field work as well as for operating stationary equipment, had to wait patiently for the development of a practical self-moving engine. While steam power was limited to portable and stationary engines on the farm, there was little hope of realizing the ambitions of those who hoped to replace horses and oxen for field work.

The problems were enormous. Early nineteenth-century steam-engines were heavy, and their modest power output was insufficient for self-propulsion under field conditions. Even when self-moving engines were developed, there was little surplus power available for pulling ploughs and cultivators. In America the situation was more hopeful, because soil conditions were often better suited for ploughing by direct traction. The heavier clay lands of much of the arable areas of Britain presented a much greater challenge.

There was no shortage of men with ideas and optimism, willing to take up the challenge. Much money, and several reputations, were lost in inventions which were sometimes too far ahead of their time, but more often, were simply impossible. The main difficulty was to devise a system of self-propulsion efficient enough for pulling implements on ground which might be hard and stoney, or wet and slippery. An additional problem was that the operating cost for tillage by steam had to compete with the cost per acre of a team of horses or oxen.

John Dumbell set out to invent a self-propelled ploughing machine, and almost invented the gas turbine. His patent, filed in 1808 in England, was for a vehicle which would be propelled by the force of steam, or possibly hot air, directed against a series of vanes in a tube or cylinder. The vehicle was to be steered by the driver, and would have several ploughs attached to the rear. The steam or air blowing against the vanes would cause them to rotate on a shaft, and this rotation was transmitted by means of gearing to the wheels of the vehicle. There is no evidence that Dumbell's invention progressed beyond the paper on which it was written.

There was much preoccupation with ropes and cables to operate tillage equipment during the first half of the nineteenth century. This line of development was later to lead to the success of the Fowler, Howard and other cable ploughing systems, where the engines were stationary and the ploughs were pulled by the cables. William and Edmund Chapman devised a cable system where the rope was stationary and the engine moved. Their idea was based on a cable or rope spanning a field, and with both ends fixed. A twist in the cable passed around a drum or pulley, which was turned by a steam-engine mounted on wheels. As the drum revolved, the vehicle literally winched itself along the cable until it came to the headland, dragging ploughs through the soil as it went. The cable could then be unfastened and moved to a new position so that the process could be repeated.

One of the most persistent of the minds concerned with ploughing by steam in the first half of the nineteenth century was the writer, J. Loudon, an Englishman who produced a whole series of ideas for harnessing steam. One of his suggestions, described in 1830, was based on a self-propelled steam-engine. He apparently realized that it would be difficult under many soil conditions to make a machine which would not only propel itself, but which would also pull a plough. To overcome this difficulty he devised an arrangement for detaching the plough from the wheeled engine. The steam-engine would be driven forward to a suitable point, and there it would be stopped and anchored to the ground. The engine power would then be applied to working a winch to wind in a cable, at the end of which was a plough. With the cable wound in, the anchors would be unfastened and the engine driven forward again, winding out the cable as it left the plough behind. This cycle would be repeated until the headland was reached, when presumably there would be complicated manoeuvres to turn the contraption for the return journey.

There were schemes for digging machines in which, it was hoped, the power put into the digging or grubbing process could be directed to pull or push the vehicle along by the reaction of the digging mechanism against the soil. Another improbable theory, which defied the power-to-weight problem of early steam-engines, was to have a belt running at one end on a steam-driven roller, and at the other end on an idler. Long, sharp spikes projected from the surface of the belt, and as these spikes clawed into the soil, the movement of the belt would force the machine to move forwards.

In a patent specification drawn up in 1811, John Blenkinsop described a form of rack railway, in which steam engines would be able to climb steep slopes by means of wheels which gripped a special notched track. His invention was designed mainly for moving materials in mining and quarrying, but Blenkinsop also envisaged a portable rack railway for providing a transport system on large farms. Five years later, in 1816, another Englishman, Joseph Reynolds, invented a 'steam carriage' for cultivations, which included an ingenious idea for turning at the headland. His machine was designed so that the wheels on one side of the 'carriage' could be reversed, while those opposite were still turning forwards. The aim was to turn the entire machine in its own length.

There were several attempts to design agricultural steam-engines which would propel themselves by means of pushers or 'crutches' which were pressed against the soil. Thomas Tindall took out a patent in 1814 for a machine of this type, which was equipped with four rods at the rear. Steam power was used to push these rods backwards against the soil. With the rods being pushed out and withdrawn again repeatedly, it was hoped that the reaction against the soil would be sufficient to move the vehicle, complete with driver and plough, across the ground. John Lee Nicolas of Clifton, Bristol, was still working on a similar theory in 1840. In that year he filed a patent for a steam vehicle which would propel itself by means of legs worked through a system of levers. He hoped the idea would be used for road transport and for agricultural work.

In France there was also some interest in the idea of a 'walking' steam tillage unit. During the early 1840s John Teissier and

Antoine Triat of Paris were working on the design of a machine which would propel itself across the ground by means of a series of crutches. The crutches moved with a walking action, operated by an arrangement of levers. These were powered by a steam-engine, which also operated spade-like diggers which cultivated the ground as the machine progressed. The idea was patented in France in 1845, but it seems unlikely that any machines were sold, owing to the practical difficulties of operating the crutches under typical field conditions.

A very different answer to the problem of achieving mobility in the field was developed and demonstrated in England by Peter Halkett, a naval officer. His *Guideway Steam Agriculture* system, patented in 1855, was based on the well-proven principle that steam-engines move more efficiently on metal rails. The Guideway was a series of metal rails laid permanently on the soil. The rails were placed 30 ft apart, which was the width of the vehicle they were to carry. This vehicle consisted of a steam-engine, followed by a small trailer or tender. The engine and trailer ran on one rail, and on one side of the engine and trailer there was a platform which spanned the 30 ft gap to the second rail. The Guideway system was later improved for larger acreages, by having two steam engines – one at each end of the suspended platform – with the extra power allowing the distance between the rails to be increased to 50 ft.

Peter Halkett's ideas had some sensible features. With the entire weight of the machine carried on metal rails, soil compaction was almost eliminated and the movement of the machine was not affected by soil conditions. The system was suitable for a wide range of arable operations, as almost any type of machine could be attached to the platform. Contemporary illustrations show the Guideway installed in a market garden situation, with a series of operations being carried out simultaneously. At one end of the platform there are containers for water or liquid manure which sprays on to several rows of plants. Near this are suspended several hoes for inter-row weeding, and next door a seeder unit drills several rows of seeds which are then covered by ridging bodies at the back of the platform. Suspended from the platform are seats or benches carrying operators who plant seedlings or weed within the rows. The trailer behind the steam-engine is laden with produce which has presumably been harvested *en route*. Other illustrations show an improbably large number of plough bodies attached to the platform.

Convinced of the advantages of the Guideway system, Halkett promoted it vigorously in public. He stressed the advantages, including accuracy of operations, and produced figures which promised cost savings of almost £1,000 on a 1,000-acre farm, or up to 40 per cent profit on the original capital outlay. To prove the efficiency of the system, he installed a Guideway on a small scale on his own farm near Wandsworth, London, using a 6-h.p. Garrett steam-engine. He later installed a second system, with a 4-h.p. Barrett and Exall engine on land he owned a few miles away at Kensington.

Halkett's ideas attracted considerable attention and some support. However the advantages, both real and imaginary, were outweighed by several considerable drawbacks. The biggest problem was the enormously high installation cost. The rails, at 30 or 50 ft intervals, would cost £20 an acre with angle-iron laid on bricks. A cheaper alternative, at half the price, would be creosoted timber rails.

Another difficulty was moving the entire contraption sideways at the end of each pair of rails. This meant more rails at the headlands arranged at right angles to the Guideway rails. The complete vehicle had to run on to the headland track, move laterally 30 or 50 ft, and then run back on to the next line of work. It was also necessary to provide rails wherever the vehicle might be needed, so that there must be a suitable track from the fields to the barns if the engine was to be used for threshing or feed preparation. Critics of the system objected to the fact that land occupied by the rails would be unproductive. In fact that rails would probably have occupied such a small proportion of a field that any loss of production would be more than offset by improved yields resulting from the absence of soil compaction.

Halkett persisted for several years with attempts to popularize

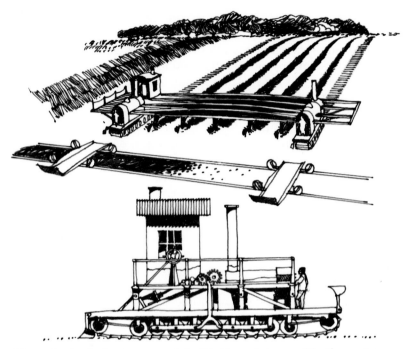

Henry Grafton's development of the 'Guideway' steam cultivation system.

his ideas, but it is doubtful if any Guideway systems were installed commercially. Although Halkett probably lost a great deal of money in the development and promotion of his ideas, he deserves some credit as a pioneer. His ideas had sufficient merit to justify further development, and there were several attempts to design out some of the faults, and particularly to reduce the installation cost.

The outstanding development from the Halkett system came from an engineer, Henry Grafton. He was probably familiar with the Guideway system, and he realized that the cost of providing rails was a barrier to acceptance of the idea. Grafton eliminated the rails, and replaced them simply and cheaply by firm, level roadways of compacted soil. The roadways were about 50 ft apart, and carried the Grafton version of Halkett's platform. Grafton planned to use two steam-engines, linked by an arrangement of metal girders or beams spanning the gap between the roadways. These beams were to serve the same purpose as Halkett's platform, which was to carry a range of implements and also to act as a rigid link between the two engines.

There is no evidence that Grafton succeeded in getting his ideas into operation, and in this respect he was less successful than Halkett. But there are two aspects of the Grafton system which now, more than a century later, seem to have been remarkably ahead of their time. One of these is the use of fixed roadways on which the vehicle travelled, and on which soil compaction was concentrated. This seems to have been well ahead of present-day thinking of planning arable operations around a so-called 'tramlines' system, so that the compaction caused by tractor wheelings is confined to a set pattern which cannot interfere with crop growth.

The second feature of Grafton's ideas was his system for propelling the device. Each steam-engine was supported on a series of small wheels arranged tandem-fashion. The small wheels were located inside an endless band, or 'caterpillar' track, and as the small wheels were driven by the steam-engine, the endless band carried the machine forward. Each of the small wheels was to be fitted with a hollow rubber tube, filled with air or water, to provide a flexible, compressible surface between the bottom of each wheel and the endless band or track. It appears that this unsuccessful invention of 1859 combined the principles of the pneumatic tyre and the tracklaying tractor.

One of the strangest farm machines to appear in Victorian England was the ploughing engine designed by Isaac and Robert Blackburn, two Derbyshire engineers. Unlike most of the more exotic inventions of the period, the Blackburn machine was built, was offered for sale commercially, price £460, and was shown publicly at the Royal Show at Salisbury, Wilts., in 1857.

The Blackburn machine consisted of a timber chassis more than 20 ft long. The chassis was supported at the front on a single wheel, which also steered the machine. At the rear the vehicle consisted mainly of an iron drum almost 10 ft in diameter and about 8 ft wide. The drum had a corrugated outer surface, and was designed to act both as the driving wheel of the machine with the corrugated surface helping to give adhesion on a wet surface, and also as a roller and clod crusher when working on cultivated ground.

As the metal drum was hollow, the Blackburns made use of the space inside to mount the boiler. The boiler, of vertical design, was suspended inside the drum on supports which were attached to the wooden chassis. There were two engines linked to the boiler, with some contemporary illustrations showing the engines mounted inside the drum, and others showing them outside the drum and mounted on the chassis. Drive from the engines was by gearing to a vast ring gear on the inner circumference of the drum.

It is likely that at least two prototypes of the Blackburn machine were built, but there are no indications of much commercial success, in spite of ambitious claims made for the machine.

It was claimed that the Blackburn cultivating machine would pull six furrows of 9 in. width at 3 m.p.h., to plough fifteen acres in a nine-hour day. With the machine operating simply as a roller or clod crusher, output was put at forty acres a day, or half that amount when pulling sets of harrows. The inventors also claimed, probably with justification, that the machine would operate with little pressure on the soil, and they stated that the compaction under the drum would be less than that of a horse's hoof. Less credible claims were made for the overall strength of the machine. It is hard to reconcile the statement that the 7-ton machine would travel on the rough public roads of the time at 6 m.p.h., with the boast that the iron drum would survive without repairs for twenty years.

The 1850s were a particularly fertile period for inventions to apply steam power to field work. Henry Holcroft announced his idea in 1856 for a machine which would literally screw itself across a field, bearing the full weight of the steam apparatus which produced its power, and also pulling ploughs or other implements.

According to John Haining and Colin Tyler in their book *Ploughing by Steam* (published 1970 by M.A.P.), Holcroft was a naval man, and his screw propulsion idea could certainly have had maritime associations. The 'screw' was a massive thread cut into the surface of a drum, which rotated by steam power. As the drum turned, the screw thread pressed against the soil surface, and was expected to propel the contraption forward.

John Haining and Colin Tyler report that a second attempt to use screw propulsion was proposed by J. A. Longden in the same year that the Holcroft idea was publicized. Longden announced that a screw thread engaging the soil surface and turned by steam power would exert a great deal of tractive effort. He claimed that his proposed machine would be most suitable for ploughing and other work where pulling power was needed.

While time, effort and money were being spent on improbable ideas, steady progress was already being made with more orthodox approaches to self-propulsion for farm work. Ransomes of Ipswich, who had created a sensation at the 1841 Royal Show with their portable engine, produced a self-propelled version the following year. This was a light-weight machine, suitable for threshing and for powering other stationary equipment, but not intended for heavy pulling work in the field. The power unit was a Davis disc engine, with drive through a chain transmission. It was mounted on a four-wheel chassis, and like the 1841 engine, it made its début at the Royal Show. Although this was not a traction engine in the proper meaning of the term, it was almost certainly the world's first self-propelled steam-engine designed for agricultural use.

Ransomes, or Ransome and May as the company was called after 1846, were also associated with another of the earliest self-propelled farm engines. This was a gear-driven engine which was announced in 1849, and was known as the 'Farmer's Engine'.

The design of this engine was very much ahead of its time. The gearing allowed two forward speeds on the road, and there were springs on the rear axle. The steering layout of the 'Farmer's Engine' was designed so that it could be operated by the man on the footplate. The earlier Ransomes self-propelled engine, and most engines produced during the next twenty years or so had the steering operated from the front of the machine, requiring a second man when the machine was on the move. Often, as with the Ransomes design, there were shafts and a horse to steer the engine, with the horse also providing supplementary power for hills or muddy roadways. The 'Farmer's Engine' had its own small trailer for carrying supplies of coal and water, and was planned as a compact, self-contained mobile power unit which could travel from farm to farm for operating stationary equipment.

Although in most respects this traction engine was of advanced design, it had no flywheel from which a belt drive could be taken. Instead there was the option of raising the rear of the machine on blocks and taking the belt drive from one of the rear wheels, or else a shaft drive could be taken off the end of the crankshaft through a form of universal joint.

The 'Farmer's Engine' was rated at just over 4 n.h.p. It weighed only 2 tons 50 cwt, and the boiler, lagged with timber slats for insulation, was only 5 ft 5 in. long. The engine was designed and built by E. B. Wilson & Co. of Leeds, but Ransomes and May were responsible for showing it at the Royal Show, which was at Norwich that year. There is some evidence that the 'Farmer's Engine' was operated commercially on several farms for threshing, and it appears to have been reasonably reliable and efficient.

Although Ransomes established an early lead in the development of self-propelled engines, there was much early competition, particularly from Clayton and Shuttleworth of Lincoln and Thomas Aveling of Rochester, Kent. While these were satisfactory as road engines for hauling light loads, they were not a solution to the problem of providing adequate power and grip for field work. The first really practical method of achieving this was to use some form of tracklaying system, and this approach achieved some success in Britain and America.

Tracklaying Engines

There are a number of competing claims to the invention of the tracklaying or 'caterpillar' principle, with patents recorded in both

France and England before 1800. The first practical attempt to design a tracklaying vehicle was probably that of Sir George Cayley, who described a steam-powered vehicle in 1825, complete with a track consisting of metal plates joined together, and running on two main driving wheels, with a number of small idler wheels or rollers between for support. This was basically the type of track which was to be used later on British army tanks, and on farm tractors particularly in America and Russia.

In America there were several attempts to make use of the tracklaying principle to develop a self-propelled steam traction engine for field operation. These date from the 1850s, when a number of patents were registered. One of the first to be built and to operate was designed by Thomas S. Minnis of Meadville, Pennsylvania and based on his patents of 1867.

The Minnis steam crawler, described by R. B. Gray in *Development of the Agricultural Tractor in the United States*, consisted of a strong timber and metal chassis carrying a vertical boiler and centrally-mounted engine. The chassis was mounted on three crawler tracks arranged in tricycle fashion, with the leading track for steering and the two rear tracks driving. The Minnis crawler was taken to Iowa for tests and demonstration, and there it pulled a plough with enough success to earn the description more recently as 'Iowa's first dirt farming tractor'.

Another tracklayer which enjoyed some success was the Stratton traction engine, produced by Charles Stratton of Moscow, Pennsylvania in 1893. The overall layout consisted of a locomotive-type horizontal boiler and front-wheel steering. The rear wheels were replaced by two tracks to provide the pulling power on arable land, and the traction engine was described by its designer as being particularly suitable for ploughing as well as for threshing and other farm work.

One of the specialized markets in North America which required purpose-built equipment, was forestry, and the advantage of steam power for the heavy work of hauling freshly cut timber to the sawmill was quickly recognized. In these conditions crawler traction engines had advantages and in the early years of the present century there were several tracklaying types of engine built for logging.

The typical design of these traction engines was a pair of rear tracks, and a horizontal boiler with a cab at the rear to shelter the driver. One of the most attractive of these was the Lombard log hauler, made about 1907. This machine had a locomotive-type layout, with the railway influence emphasized by having a saddle tank over the boiler. The Lombard, which was described as 100 h.p., had a pair of skids at the front, instead of wheels. Alvin Lombard, who designed the log hauler, had been experimenting with crawler-type traction engines for forestry work in New England since the 1890s, and some of his earlier designs were among the first commercially successful tracklaying machines in the world.

Another of the specialized markets in the United States was for equipment to cultivate the fertile peatlands of California. This was an area of prosperous, large farms, requiring tillage equipment to cope with ground conditions which were often soft. The tracklaying system met the requirements, and two companies competed vigorously to supply crawler-mounted steam-engines in the area. The two firms were Holt and Best, which after years of rivalry merged, and very much later became the Caterpillar Company specializing in farm and industrial tractors, and noted particularly for their crawler tractors. Both the Holt and the Best companies were building wheeled traction engines during the 1890s, and both faced the problem of designing wheels which would support the weight of the steam-engine without sinking into the soft peat soil. One short-lived solution was to make the driving wheels so wide that there would be a big surface area of peat to support the weight. In 1901 the Best company announced a traction engine with wheels 15 ft wide on each side, and two years later Holt designers produced a vast machine with 18 ft of wheel width on each side. In the San Joaquin Valley this traction engine cultivated and drilled in a 44 ft wide strip.

Although the crawler track was largely a British invention, engineers in the middle of the last century in Britain became side-tracked by an alternative system developed by James Boydell. This

An alternative to the Boydell 'Railway' was the Cambridge system, patented in 1856, but based on the same principle.

tracklaying or 'railway' idea was used experimentally and commercially from about 1854. Boydell preferred to encourage established manufacturers of traction engines to fit his 'railway' tracks, instead of building complete engines himself. About six companies used the idea, and some of these, including Tuxford of Skirbeck, Lincolnshire and Burrells of Thetford, Norfolk, each built a dozen or more engines with Boydell wheels. Some of the leading farmers and commentators of the 1850s approved of the Boydell principle, and Burrell–Boydell engines were exported in significant quantities.

The Boydell 'railway' consisted of a number, usually five, of flat plates loosely hinged around the circumference of the driving wheels. As the wheels turned, the hinged flaps automatically came down in succession, to lie flat on the ground to give a continuous roadway for the wheels to roll on. Some of the engines had three or four wheels equipped with the Boydell apparatus. In theory, and to some extent in practice, the Boydell 'railway' was a success. The flat plates or flaps certainly helped the wheels to roll over soft or uneven ground, and to grip for ploughing and cultivating. Under test the Boydell system allowed traction engines to exert more drawbar pull than conventional wheeled engines could exert at that time. The idea was attractive enough to encourage several patents for imitating or improving the basic Boydell idea.

In spite of some genuine merit and a great deal of promotion and demonstration, the Boydell system faded from popularity in the mid-1860s, only ten years after it was first used. Boydell failed to make his fortune, or even to recover his development and promotion costs, because the 'railway' proved too unreliable and expensive to maintain. The hinges on which the flat plates were supported, wore rapidly in abrasive conditions of sandy or stoney land. Big stones damaged the plates, adding to the maintenance cost. The Boydell system was also criticized for causing soil compaction – which one would have expected to be less than with a conventional traction engine – and the initial cost was high.

4

STIRRING THE SOIL

Ploughing is a slow, and usually expensive form of cultivation. This is true now when there are tractors to pull the plough, and it was also true when the plough was pulled by horses or oxen and when an acre a day was a reasonable output for a team. Now, with the development of chemicals to control weeds, there is a significant move away from the mouldboard plough to some form of reduced cultivation. In the nineteenth century there were many who saw the power of the steam-engine as the means to make a break from the plough. These were the farmers and engineers who tried with great determination to devise some form of powered cultivator, usually with a rotating action, which would make the mouldboard plough redundant.

Among those in Britain who sought to replace the plough, was the *Farmers' Magazine*, according to editorial comments published in 1855. The comments were made in connection with a prize offered by the Royal Agricultural Society of England for a steam-powered cultivator, and included a list of objectives for a machine which would replace the plough. The six points were:

1. To turn the soil as perfectly as a spade and better than a plough.
2. To work at variable depth on any soil type, and cause no more soil compaction than a horse's hoof.
3. To work across furrows and to cope with weeds.
4. To deal properly with grassland and invert it.
5. Be convenient to use and without complications.
6. The cost must be no more than that of the horses and implements of equal work output, which the new steam machine would replace.

Throughout the period in which steam power was popular on farms, nothing appeared to match the virtues demanded by the *Farmers' Magazine* editorial. This was not for lack of effort, and there was a widespread search, in Britain, the U.S.A., Canada and Europe, for a steam-powered machine which would outdate the plough. British inventors, as usual, made the most determined effort, but there were notable contributions from overseas.

While many of the rotary mechanisms imitated the action of the spade or the plough, the machine designed by Pierre Barrat in 1847 attempted to reproduce the action of a mattock. Barrat, a Frenchman, mounted his steam-powered mattocks on a framework at the rear of a self-propelled four-wheel engine. The complicated motion of the mattocks was partly to-and-fro, and partly a circular movement.

Another French contribution was a steam-powered cultivator designed by M. Kienzy. This machine appeared in 1863 at an international exhibition at Lille in France, where it competed against a Howard steam plough from Britain. British steam equipment generally dominated the awards lists at the Lille event, but the Kienzy machine attracted unusually favourable comment from British observers. In their report in the *R.A.S.E. Journal* for 1863, they referred to the Kienzy machine as being superior to steam cultivators seen in Britain. Set to work in heavy, couch-infested soil, M. Kienzy was able to operate his machine without difficulty to a depth of almost 14 inches, and the result was a fine tilth.

The Kienzy machine had an 8-h.p. engine turning at 60 r.p.m., with the boiler working at 80 to 90 lb pressure. The digging mechanism was powered through a series of rods and chains, and consisted of horseshoe-shaped forks on a framework at the rear of the engine. The working width was 5 ft 8 in., and the output was said to be half an acre an hour. British observers were impressed not only by the quality of the work, but also by the manoeuvrability of the machine. This was said to be achieved by some form of differential mechanism.

One of the most imaginative or optimistic of the nineteenth-century inventors working with steam power for farming was the

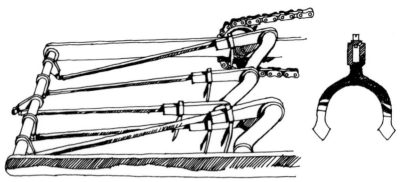

Diagram of the Kienzy steam cultivator mechanism.

Canadian, Robert Romaine. His home was stated vaguely in one British publication of the time as 'Peterborough, West Canada'. It is said that Romaine received financial assistance with his inventions from the Canadian Government, but his reputation was greatest in Britain and France.

The first of Romaine's inventions to attract attention was a rotary cultivator which was powered by steam but designed to be pulled in the field by horses. This arrangement was quite sensible in the early 1850s, when much effort was being wasted in vain attempts to design a self-propelled tillage machine. Another feature of this machine, which was patented in 1853, was that the rotor could be removed and a cutting mechanism fitted, so that the whole machine could be operated as a steam-powered reaper. So that the horses would not trample the uncut crop, in the harvesting version the machine was to be pushed rather than pulled.

As a rotary cultivator, there was provision in the design for spikes, spading bodies or forks to be fitted to the revolving cylinder, so that the work done could be varied. Romaine planned to fit a roller behind the rotary cultivator, and between the cultivator and the roller there was to be a seedbox and seeding mechanism. A levelling board to level the seedbed surface could also be fitted.

Romaine's first machine was too far ahead of farming technology and probably also ahead of engineering capability in 1853. However, he deserves recognition for his attempt to invent a complete single-pass minimum cultivation drill, to cultivate, seed and roll in one operation, with a machine which would later be used to harvest the crop.

Not surprisingly, Romaine attracted the attention and support of notables such as Mecchi and a Yorkshire engineer and machinery manufacturer named Alfred Crosskill. After his first machine proved impractical, Romaine, who was then working in Montreal, built a self-propelled rotary cultivator. This was patented in London in 1855, and a model was exhibited in the same year in Paris. Two of the self-propelled version were manufactured in Crosskill's works in Beverley, and earned favourable reports from farmers in Yorkshire who had work done by the machines.

Samuel Copland, in his book *Agriculture Ancient and Modern* published in London in 1866, devoted several paragraphs to Romaine's rotary cultivators. He reported that the self-propelled machines had failed because of their great weight. At 15 tons, they made ruts even on established clover leys which had been baked hard by the sun. When Copland was writing his book, however, Romaine was still working to perfect his steam-powered rotary cultivator – this time in a version which could be pulled like a plough on a cable.

John Algernon Clark, writing in the *Royal Agricultural Society Journal* in 1859, suggested that the Crosskill–Romaine machines weighed 10 tons, and confirmed the good reports on the performance. Working rate, he reported, was between 4 and 7 acres a day at a cost of between 5s. and 10s. (25p and 50p) an acre. He reported that Romaine's machine had been improved in a simplified version built by W. H. Nash of Cubitt Town, Isle of Dogs, London.

American inventors were also intrigued by the possibilities of using steam to stir the soil, and ideas for rotary cultivators were numerous in the second half of the nineteenth century. Elisha Otis, remembered for his connections with lifts or elevators, took out a patent in 1857 for a rotary steam plough. A year later Henry Platt of New York filed his patent for an ingenious rotary plough, to be made in small sizes for horses to pull, and larger models for towing behind a traction engine. The rotary action was transmitted through gearing from the wheels of the plough. Samuel Reynolds and Elias Howe both published ideas for rotary cultivation with steam. Reynolds is better known as the inventor of machinery for making nails, and Howe is described as the inventor of the sewing-machine.

The most successful American design for a steam-powered rotary cultivator came from Martinez, California, where Philander Standish demonstrated his brainchild in 1868. This machine had a centrally mounted vertical boiler and two horizontal cylinders. The rear driving wheels were each 2 ft wide in order to provide traction on soft ground. The rotary cultivator mechanism was powered by the engine, and consisted of a varying number of horizontally rotating sets of tines. Depth of work was adjustable and governed by a depth control wheel. Reynold Wik in his book *Steam Power on the American Farm* (University of Pennsylvania, 1953), records that Standish spent £2,500 developing the machine, and he used it in 1868 to cultivate 100 acres in the Diablo Valley. In spite of this evidence that the machine was capable of working, it was not a commercial success.

With so much effort being invested in nineteenth century efforts to produce a practical rotary cultivator, it seems remarkable that the progress made was so slow and so slight. Dan Pidgeon, writing in the 1892 *Journal of the Royal Agricultural Society*, admitted that there were at that time some slight indications of commercial success. But the earlier obsession with this idea had proved harmful, he suggested.

'There is no question but that the introduction of steam-ploughing proper was long retarded by the firm hold which the idea of "rotary" cultivation had taken on the public mind,' he wrote. 'That "digging" has still an extraordinary fascination for certain inventors is shown by the persistent appearance at agricultural shows of steam digging-machines. These implements, ingenious and mechanically meritorious as many of them are, have, as yet, accomplished little.'

48

Among those in Britain who contributed to the rotary tillage obsession were such familiar names as Richard Trevithick, one of the outstanding engineers in the history of steam-engine development, and James Usher, famous as a successful Edinburgh brewer.

Trevithick's contribution to rotary tillage was the design for a machine which he called the 'steam spade-tormentor'. Drawings for this machine were produced in 1813. The spade-tormentor was designed to be pulled by a steam-engine – possibly by direct traction or else by a rope from a windlass. A wooden framework mounted on four wheels, carried a series of tines for breaking the soil. At the rear of the vehicle was a large wheel, mounted vertically, and made to rotate by gearing from the land wheels. The rotating wheel carried a series of spade-like digging bodies around the circumference; and these stirred or dug the earth already loosened by the tines.

Usher took time off from his brewing business to design a steam-driven rotary plough, which he patented in 1849. A prototype version was built by an Edinburgh engineering firm, and was ready for trials at Niddry Mains, near Edinburgh, in the spring of 1851. This machine consisted of a 10-h.p. engine mounted on a four-wheel chassis. The rotary plough bodies were carried on a framework at the rear, and were power-driven to rotate in the direction of travel.

The trials, carried out in public, were impressive, according to contemporary accounts. The machine had a forward speed of 3 m.p.h., to give a theoretical output of 9 acres a day. In practice, allowing for stops and turns, the commercial output was rated as 7 acres a day. Costs were estimated, probably with a degree of optimism, at 17s. 6d. a day, and the machine was shown to be capable of climbing a 1 in 10 gradient. Encouraged by these results, Usher put more money into the project, and a second, improved, machine was completed in 1855. While the prototype had used a horizontal boiler, the Mark 2 version was designed with a vertical boiler and double cylinders. A very advanced feature was a power-operated lifting mechanism to raise the cultivating unit out of work.

Presumably confident of success, Usher took his new machine to the Royal Show, which in that year was at Carlisle. The Royal Agricultural Society had a field set aside for a trial of steam-ploughing equipment, but for some reason the Usher steam plough could not be moved there.

So that the trials could proceed, the judges sought, and gained, special permission from the local council for the steam plough to be used within the showground itself. This was an area of ground belonging to the council, consisting of flat, easy working, deep soil with permanent pasture. The machine was set to work, but the judges' report suggests that the results were disastrous.

'The machine, under the most favourable circumstances, could with difficulty move itself, and the revolving shares neither inverted nor pulverised the soil, but tumbled it about in wild confusion, and left it in a state more unfavourable for cultivation than it was before', said the official report of the trials. After this public disaster, the Usher steam plough more or less vanished from the scene. Its designer earned little reward for his investment of money and time in a machine which was an outstanding attempt to contribute to farm mechanization.

Paul Hodge, of the Adelphi, London, also earned little reward for his contribution to technology. He designed a steam-powered digging machine which was supposed to have a special application for digging organic manures into soil. Steam power was used to press one or more 'spades' into the soil, while at the same time compressing a spring. At a suitable stage in the digging cycle, the downward pressure was released, allowing the spring to flick the spades with their loads of soil, upwards and outwards. This was intended to dig and break up the soil, and at the same time to mix in the manure.

A more successful digging machine appeared in about 1876, designed principally for digging hopyards. The inventor was John Knight of Farnham, Surrey, a hop grower who turned to mechanized digging because 'agitation among the labourers' had raised the cost of doing the job by hand. The digger was operated by a stationary or portable engine which powered a cable or rope. The

Garrood's rotary digger of 1892 – one of the many unsuccessful
nineteenth-century designs.

movement of the cable acted on a pulley on the digging machine,
moving it forward and actuating the digging forks.

Mr Knight put his machine on the market, priced at about £140,
plus, of course, the cost of the steam-engine. It appears to have
proved a commercial success, because the patents were bought by
Howard of Bedford, who added the hop digger to their successful
range of steam-ploughing tackle.

The long list of names of would-be rotary tillage inventors,
includes a few more who achieved modest success before the end
of the century. Success came in the form of rotary digging mechanisms which were supplied more or less as attachments for standard
traction engines. Three names were outstanding in this type of
cultivation – Cooper, Darby and Proctor. The first of these to

appear was the Darby 'pedestrian broadside digger', which was the prototype for further development. Thomas Darby, from Chelmsford, Essex, showed his machine at the 1878 Smithfield Show in London. It was a wonderful contraption, weighing $12\frac{1}{2}$ tons, 18 ft wide, and propelled forward by an arrangement of steam-operated legs.

From this somewhat crude beginning, Darby progressed towards more compact designs. These resulted in 1891 in a 'Quick Speed Digger', designed as an attachment for a standard traction engine, which would require little modification to accept the digging mechanism. In the same year the Darby Land Digger Syndicate Ltd, was granted an extension to their patent cover, in order to provide further time in which to develop the idea commercially. Although the digging attachment was more versatile and more successful than the original form of a complete machine, the company never achieved a really profitable basis.

Thomas Cooper started making steam-powered diggers in 1891 at King's Lynn, Norfolk. These started as complete machines, but later the Cooper digging mechanism was available for attachment to various makes of traction engine, including Fowler. A feature of some Cooper machines and conversions was that a single front wheel was used to aid manoeuvrability at the headlands. Cooper achieved a good deal of success with export business, and also some success at home against Darby machines. Trials for steam-powered diggers were held by the Royal Agricultural Society at York in 1900, with Cooper competing against Darby. The Cooper machine gained through better manoeuvrability. Although the Darby digger had a greater working width and an equal forward speed, it also had a greater turning circle. While the Darby machine turned in 82 ft, the Cooper turning circle was less than 25 ft. The result was a victory for the Cooper machine.

The third of the trio of late nineteenth-century steam diggers was the design of Frank Proctor of Stevenage. The first machines produced for Proctor were exported to Spain, and these were completely built to his design.

Later, he collaborated with Burrells of Thetford to produce an attachment for their traction engines, and this was a commercial success. In 1882, in a demonstration arranged for a Turkish Government official on a buying mission to Britain, a Proctor conversion on a 10-h.p. traction engine worked at the rate of 7 acres in 10 hours. The cost of fuel and labour was calculated at 4*s.* an acre, including an allowance for depreciation and maintenance, and based on 180 working days a year.

The 4-ft wide forks of the Proctor machine were actuated by cranks, and were designed to work at 45 revolutions a minute. With a forward speed of 45 ft per minute, the forks were supposed to dig at one-foot intervals. The Proctor machines were particularly successful in overseas markets, and were also used by contractors in Britain, who charged 10*s.* an acre, or up to 15*s.* an acre for deeper work.

5
GOD SPEED THE PLOUGH

While the search for a practical system of rotary cultivation diverted some enthusiasts, there was still plenty of effort, imagination and cash to be invested in the search for a system of ploughing by steam.

To encourage the search, substantial cash prizes were offered in Britain for a successful steam-ploughing system. Some of the cash offered was made available by wealthy landowners, but the leaders in offering incentives were the Royal Agricultural Society of England and the Highland and Agricultural Society of Scotland. In addition to these incentives, would-be inventors were also encouraged by the prospect of big commercial rewards to be earned. It was widely believed that the steam plough would achieve a great breakthrough allowing faster, deeper, cleaner ploughing, and at a lower cost than for ploughing by horses or oxen.

During the middle decades of the last century there was plenty of verbal support for the idea of ploughing by steam. The subject was argued at length and with enthusiasm from the platforms at meetings, and there were frequent references in the press. Pamphlets promoting the idea were published in Britain, and steam ploughing was also referred to in a popular novel, *The Mummy*, published in London in 1827 and written by J. Loudon, a well-known advocate of steam cultivation.

One of the outstanding pioneers of steam ploughing was J. A. Williams, who experimented with rope or cable systems on his own farm at Baydon, Wiltshire. His practical experience earned him the privilege of speaking at a meeting of the Farmers' Club in London in May 1855, when he offered his audience some indications of the potential value of steam tillage.

John Williams claimed that steam tillage, completely adopted on British farms, would release four horses for every 100 arable acres.

With 47 million acres then cultivated in Britain, 1,880,000 horses would be replaced. The average value of these horses was £30 a head, thus farming would no longer require £56,400,000 worth of horses. Each of these horses, Mr Williams explained, cost £30 a year to keep, making a further saving of £56,400,000 a year when they were replaced by steam. Part of the cost saving would be in food, which would now be available to produce 80,571,000 stones of meat, 805,715 fat cattle or 8,057,150 fat sheep.

Speakers at a Smithfield Show meeting in London in 1863 were equally enthusiastic about the cash rewards of steam tillage. There was a general anticipation that the cost of ploughing by horses would be reduced by steam power to save 30 to 40 per cent, and there would be an extra bonus in the form of increased crop yields. One speaker predicted that the greater yields to be gained by using steam in the field, would mean a direct annual saving to the nation of £20 million worth of grain imported from eastern Europe and North America. With so much shipping space no longer needed for our grain imports, the cost of sea freight would be forced down, saving the nation a substantial amount on our general imports from abroad.

The benefits of replacing the horse for ploughing would not be simply a matter of cheaper tillage and better yields. The Smithfield Show audience was promised that even fox hunting would be improved, as steam power would allow deeper ploughing and better drainage. Sir George Jenkinson, speaking at the meeting, explained how much easier it would be to hunt over heavy clay land which had been improved by steam cultivation. And the land would suffer less from the horses' hooves, he thought.

It was even thought that the steam plough would be good for the farm workers' soul. The Rev. H. W. Beecher, writing in 1855 for the *Independent*, presented his views on the coming of the steam plough. 'The man who invents a steam plough that will turn twelve or fifteen acres a day, two feet deep, will be an emancipator and civilizer,' he wrote.

'Then Labour shall have leisure for culture. Thus working and studying shall go hand in hand. Then the farmer shall no longer be a drudge; and work shall not exact much and give but little. Then men will receive a collegiate education to fit them for the farm, as they now do for the pulpit and the forum; and in the intervals of labour, gratefully frequent, they may pursue their studies; especially will books be no longer the product of cities, but come fresh and glowing from nature, from unlooped men, whose side branches having had room to grow, give the full and noble proportions of manhood from top to bottom. God speed the plough.'

With so much to commend it, the steam plough was slow to arrive in Britain. Meanwhile there were rumours, presumably ill-founded, that faster progress was being made abroad.

A claim that a French inventor was ahead of British research was published in the *Gardener's Gazette* in 1838 in a report on a new 'steam plough' which had been demonstrated in France. According to the rumour, this machine could dig a trench 2 ft wide, 1 ft deep and a mile long in one hour. It appears that the mechanism of the machine lifted soil against a 'sail' which then threw the soil aside. The machine was steam powered, but no other details were quoted in the brief report.

Even more formidable was the report appearing in Britain of a story which originated apparently in the *Philadelphia Gazette* in 1838. According to this rumour, a Mr Campbell of Philadelphia had invented a steam-powered machine which would plough or harrow 250 acres a day. 'If this is the case,' the *Philadelphia Gazette* story was quoted as saying, 'The wilderness of our great prairies will be made to blossom like the rose.'

While commentators in the British press were seemingly eager to publish rumours and opinions about a subject as topical as steam tillage, they seem to have been less diligent in recording when the first success was achieved. There is uncertainty about the results of some of the patent applications recorded during the first thirty years of the last century, and there are also several suggestions – but without proper detail or evidence – that some ploughing was done with steam power several years before the widely recorded achievements of John Heathcoat. One unconfirmed report refers to ploughing in Lincolnshire in 1836, with the plough operated by means of a cable from a portable steam-engine on the headland. Another report indicates that a similar system was used a year earlier than that, and we also have a report of mole ploughing with a cable system, demonstrated publicly in 1829 at the West Kent Agricultural Association ploughing match.

With insufficient evidence available about earlier achievements, most of the credit for being first in the field is usually awarded to John Heathcoat. There is some justice in this, for although he was almost certainly not the first, he was among the most notable of the pioneers who achieved anything at all in public. While others were attempting to adapt existing portable engines for ploughing with some form of rope or cable, Heathcoat chose the much more difficult and more costly approach of using a specially designed vehicle. He also deserves some recognition for choosing to plough the most challenging undrained marshland, rather than a field of easy arable loam.

John Heathcoat, M.P. for Tiverton, Devonshire, started his career as a textile manufacturer, and his engineering achievements include more than twenty patents for machinery for the textile industry. Later he became interested in the problem of improving undrained land, and at this stage he worked in partnership with Josiah Parkes, an engineer with a special interest in land drainage problems. Heathcoat and Parkes took out patents in 1832 for their steam-ploughing apparatus, and after development work on the prototype, were ready for extensive trials and demonstrations. These took place in 1835 on an area of boggy land known as Red Moss, near Bolton, Lancashire.

The power source for the Heathcoat and Parkes ploughing system was a 20-h.p. 2-cylinder steam-engine. This was mounted inside an extraordinary contraption, with four sets of three wheels each – all 8 ft in diameter. These sets of wheels carried two endless tracks, which were made of timber with metal joints. Between the front and rear sets of main wheels on each side were a number of small idler wheels to support the track, making this one of the first tracklaying machines to operate.

The front and rear axles of the Heathcoat and Parkes machine supported a platform on which the engine and boiler were located. The engine provided power for two winding drums, each capable of being reversed. The drums were used to wind in a metal strap, which acted as the cable to pull one plough for each drum. The machine was topped by a fully pitched roof – at least in a later version of which pictures survive – giving a height of nearly 12 ft. Total weight was about 30 tons, and the crawler-type tracks were intended to carry this weight over soft marshland.

Heathcoat designed the machine to work with one plough at each side. The strap passed around a movable anchor at the end of the furrow on each side, and this could give a working width of up to 220 yards both sides. The two drums were designed to work alternately, with one winding a plough in while the other plough was being pulled out towards the anchor. After each bout with the ploughs, the engine and drums had to move forward by the amount of the furrow width, ready for the new bout. Although there are suggestions that the engine propelled the vehicle forwards, it seems likely that this could only be achieved by winching the machine forward, or perhaps by using horses. The ploughing system, with land ploughed at both sides, meant that a central strip, along which the tracked machine moved, remained uncultivated. There was even a patent prepared for a self-propelled roller to precede the ploughing machine in order to flatten the ground for it.

The ploughs were almost as unusual as the rest of the ploughing system. Each plough had scissor-like knives operating at the front to chop through tangled vegetation and roots, in order to clear a passage for the furrows. The theory behind the whole operation was that by ploughing areas of undrained bog, surface drainage would be improved and the cultivated soil would carry more productive herbage and support livestock.

The Red Moss trials were sufficiently impressive to attract widespread interest and some favourable comment. Arrangements were made to transport the complete ploughing system to an even more difficult area of marshland in Dumfriesshire, Scotland, where a public demonstration was arranged in 1837. These trials, in conjunction with the Highland and Agricultural Society, were a bid by Heathcoat to win a £500 award offered by the Society to the first person to demonstrate a practical system of steam cultivation.

After several days of mixed success and failure, Heathcoat and Parkes ended the trials disappointed. The Society decided to award only £100, in recognition of a good, but unsuccessful, try. Although the system actually achieved some ploughing, there were mechanical failures, particularly of the metal straps pulling the ploughs. The trials also showed how small the output was in relation to the vast cost of the machine, and also in relation to the team of seven men needed to work it. There was doubt about the economics and also about the agricultural value of simply ploughing the surface of marshland.

Their lack of success must have been a bitter disappointment to the partners, and especially to Heathcoat. He had invested at least £12,000 of his own money in the development work, and he had also given a good deal of his valuable time to the project. The postscript to the Heathcoat adventure in Scotland, according to some reports, was that the immense ploughing engine was abandoned after the trials, to sink beneath the surface of the bog. One possible explanation is that the cost of moving the machine from the marshland and transporting it back to England, was more than the partners were prepared to pay after their disappointment. It is not clear whether the machine actually sunk out of sight, or if it merely settled into the mud, to rust and rot away, a convenient source of scrap metal for anyone who cared to claim it.

The Heathcoat and Parkes ploughing system made use of several ideas which were by no means new. The idea of using crawler tracks had already been patented, and this applied also to the system of using rope or a cable to transmit steam power to pull a plough. The cable ploughing system, which was to dominate steam tillage in Britain, appears to date back to patents taken out by a Middlesex farmer, Major Pratt. His patent was filed in 1810, and refers to the use of a chain passing around a system of pulleys. The chain could be used to pull various implements, including a plough, and its movement could be powered by steam, or possibly also by wind. There is no evidence that Major Pratt's patents were ever proved by him in practice. If he was able to make his ploughing system work, he would deserve very much more recognition than he has so far been given, and certainly more than John Heathcoat. However, the chances are that his ideas progressed no further than the patent, for in 1810 the technology of steam-engine and boiler design was at an early stage to cope with what would have been an inefficient means of transmitting power, and certainly his alternative suggestion of using wind power must have been a non-starter.

Ropes and Cables

Although Major Pratt may have failed to make his chain-ploughing system work, and Heathcoat achieved only partial success with his attempts to use a somewhat similar principle, the idea of using some form of rope, cable, strap or chain was beginning to attract great interest. Those who worked on ideas of using direct traction methods to plough behind a self-propelled steam vehicle, were in the minority in Britain. This generalization includes John Upton who announced a direct traction system for ploughing, with the added refinement of using a rotary steam-engine instead of the piston-engine, which was advanced thinking in 1837.

Another pioneer who could reasonably claim the distinction of developing the first functional steam-ploughing system was a Scotsman, Alexander MacRae. He designed an ingenious arrangement of a cable running between a steam-engine and an anchor, with the plough being moved to and fro between the engine and the anchor. The odd feature of his ploughing system was that it was devised specifically for British Guiana, where MacRae was a sugar planter. In the Demerara region sugar was cultivated on low-lying, level land, which had been intersected by canals dug for irrigation purposes and arranged in a pattern of parallel lines. MacRae decided to put his steam-engine and winding drum in a barge, which could move slowly along a canal, with the anchor in a second barge moving equally slowly along a parallel canal. The cable would carry the plough across the ground between the canals.

Perhaps, surprisingly, the MacRae system was actually set up on a sugar estate, where it was said to be working efficiently with a three-furrow plough. This system was patented in England in 1839.

A significant development of the MacRae system was patented in 1846 by John Osborn, who was also connected with the sugar estates of British Guiana. John Osborn used the idea of a rope or cable to pull a plough, and he also relied on the system of parallel canals to float barges. However he progressed by deciding to use two steam-engines, each fitted with a pair of winding drums, to operate two ploughs simultaneously. He was able to make his ploughing system more versatile than that of MacRae by suggesting that the two ploughing engines might run along parallel sets of rails laid along the sides of a field, as an alternative to the barges for British Guiana conditions. Osborn was able to demonstrate his ploughing system successfully in Britain as well as in British Guiana. Arrangements were made to manufacture the equipment in Britain on a modest commercial basis, and the results achieved probably did much to concentrate attention in Britain on cable ploughing rather than direct traction. According to Mr Harold Bonnett in his book, *Saga of the Steam Plough*, Osborn attracted the interest and support of the Marquis of Tweeddale, who became one of the leading advocates of steam cultivation in Scotland. The Marquis ordered a twin-engine set of ploughing tackle to be built for his estate in East Lothian, and this equipment was based on Osborn's ideas.

Another member of the British aristocracy to give practical support to steam-ploughing development was Lord Willoughby d'Eresby, the owner of a large estate in Lincolnshire. He chose to base his trials on a cable system, or rather a chain, and started using the equipment on his land in 1850.

Several engines were used during Lord Willoughby's period of experimenting with steam ploughing. The first, and the best known, of these was a portable with a nominal rating of 26 h.p. This engine was constructed at the engineering works in Swindon, Wiltshire, where locomotives for the Great Western Railway were built. The man who designed this engine was probably Sir Daniel

Lord Willoughby d'Eresby's first system of steam ploughing.

Gooch, one of the most famous of the railway engineers in Victorian England.

Lord Willoughby's first arrangement for steam ploughing was to have the engine and winch mounted on a wooden platform, which could be pulled along a temporary railway track. The wooden rails were laid across the middle of a field due for ploughing, and were moved elsewhere after the engine had passed. As the engine passed over this track, it provided power for two ploughs. These ploughs were designed to operate at the ends of a single length of chain, with one plough working at each side of the engine. The chain was looped around the capstan drum of the winch, and as the capstan was reversible, it could pull in the chain from either side. As one plough was pulled in towards the engine, turning a furrow as it went, the other plough on the slack end of the chain was pulled by a horse away from the engine and towards the headland. The plough moving outwards would be lifted out of work, and could be moved easily by one horse. When the incoming plough reached the capstan, and the outgoing plough reached the headland, the arrangement had to be reversed. Both ploughs would be turned, the horse unhitched from the plough at the headland, another horse hitched to the plough by the capstan, and the capstan would be reversed. The plough at the headland would travel towards the capstan turning a furrow as it went, and the other plough would be pulled away empty by a horse, taking out the slack chain as it went.

So far his Lordship had devised little more than an expensive system for making life easier for his horses. The two ploughs were both travelling empty for half of their journeys, and a substantial amount of time would be spent at the end of each furrow hitching or unhitching horses. Frequently there would also be a pause to allow the engine to be winched forward to keep the ploughs on unworked ground. The hardest work for the horses would be to plough the odd corners which the steam ploughs could not reach, and also to plough the uncultivated path down the centre of the field where the engine and winch had been carried on the railway track.

The system was greatly improved by having movable anchors at the headland on either side. With a longer, continuous-chain, the ploughs could be moved out to the headlands by engine power instead of by horses. In a later development of the system, which was demonstrated publicly in the spring of 1852, two engines were used, one on each headland. This meant that no time was wasted with the plough travelling 'empty', and there was no central strip to be ploughed by horses afterwards. Lord Willoughby announced that visitors were always welcome to see his steam-ploughing apparatus in action, and he also announced that others could make free use of his ideas if they wished.

Although a practical system of ploughing with portable engines was obviously emerging, there were still those who preferred to try to plough from a stationary engine. Augustus Lacy of Knayton, Thirsk, Yorkshire announced his scheme in 1856, and it was described in *The Engineer*, 2 May edition for that year, complete with a drawing of a plough working at 3 m.p.h. – or so the caption claimed.

Augustus Lacy explained his ploughing system in terms of a 400-acre arable farm, but it would have been adaptable for other farm sizes. His power source for 400 acres was seven stationary steam-engines, all to be permanently sited. Six of these engines would be of 8 h.p. each, and these would be located so that with convenient lengths of rope or cable they would plough all but the central area of the 400-acre block. The central area would be ploughed with power from the seventh engine, which was to be 12 h.p., and with capstans to enable it to plough in all directions. The boundary of the 400 acres in the Lacy scheme, was to be an iron fence, sturdy enough to be used throughout its length as the anchor point for the cables pulling the ploughs.

The inventor of the system agreed that it might appear to be extravagant in the use of steam-engines and fencing, and he also agreed that the system would not work anywhere but on level or evenly sloping land. The advantage, he claimed, was that the cable system from each engine could be simple and mechanically efficient. Augustus Lacy also designed a special plough for deep working, which was one of a range of implements his steam system could operate. This plough had two furrows arranged one above the other, with the top mouldboard taking the upper slice of the soil, and the lower mouldboard taking a further slice for really deep work. The two mouldboards were arranged so that each layer of soil lifted was inverted, and also interchanged – with the former top soil being dropped under the original deep soil. Each plough had a seat for the operator, who could communicate with the engine attendant by means of 'Gluckman's patent electric signal'.

In spite of the progress which had been made and demonstrated publicly by the mid-1850s, there were still many who believed it would be impossible to use the power of steam to pull a plough. This widely held belief is indicated by the report of a meeting held in London by the Farmers' Club in June 1853.

The principal speaker for this meeting was Alan Ransome, and his paper dealt with the place of portable and stationary steam-engines in British agriculture. Although there was very wide agreement that both types of engine had great value on the farm, there was equally wide agreement that steam power could not be adapted to ploughing. In fact there was some amusement in the audience at the mere suggestion that men might plough with the aid of a steam-engine.

Such a pessimistic attitude was already being proved wrong. The pioneering work of the first half of the nineteenth century led to the development of rope or cable systems which became commercially acceptable in the second half of the century.

A major breakthrough came in 1850 when a Berkshire farmer by the name of Hannam developed what was later to become known as the 'Roundabout' system. This made use of a single steam-engine – usually a portable for which there were other jobs on the farm – together with a very long continuous rope or cable and a windlass. The rope or cable was laid around the edge of the field to be ploughed, fixed at the corners of the field by anchors and supported at intervals along the sides by 'porters'. The porters allowed the rope to run freely, held about 2 ft above the ground. The rope or cable was kept running around the circuit by means of the windlass, which was stationed with the steam-engine at some convenient point by the edge of the field.

The 'Roundabout' worked simply by using the motion of the rope or cable to pull a plough or other implement. The line of travel of the plough was adjusted by moving the anchors, and the position of both porters and anchors could be arranged to cope with fields of various shapes. The system had much to commend it, with a particular attraction being the relatively low capital cost. The power source could be a portable engine, of which there were an estimated 8,000 on British farms in 1851. This meant that the same engine could be used for a wide range of other jobs to help spread the cost. Existing ploughs or cultivators could often be adapted for use with the system, and for working on wet, heavy soils, there was the further advantage that the weight of the engine did not go on to the ground being cultivated, to cause compaction problems.

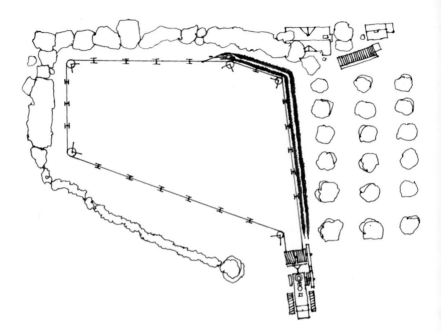

Diagram of the William Smith 'Roundabout' cable system.

Mr Hannám had his windlass and other equipment especially built for his farm, and he appears to have ignored the possible opportunity to develop the system commercially. This was done to some extent by another farmer, William Smith of Little Woolston, Buckinghamshire, who developed a form of roundabout system, but with some improvements, which he used on his own farm in 1853.

One of the refinements added by Smith was the idea of making the direction of travel of the rope or cable reversible. His windlass could be made to work in both directions so that a plough could be moved to and fro. In its original form the William Smith roundabout was designed to work with a special heavy-duty tined cultivator of Smith's own design. A roundabout system was later marketed by the brothers J. and F. Howard of Bedford, using Smith's ideas, and this version was available with ploughs. The Howard roundabout system was reasonably successful in Britain, with the relatively low output compensated for by a lower price than some competitive equipment.

A variation on the roundabout theme was shown publicly in 1852 by the three Fisken brothers from Northumberland. Credit for the idea is usually attributed to brother William, who was a minister in the Presbyterian church. Although the individual contributions made by each of the brothers may not be certain, there is no doubt that a contribution in the form of financial assistance to meet some of the development costs came from the Highland and Agricultural Society of Scotland.

The Fisken system transmitted power to the plough through a rope, as with the Smith system. The special feature of the Fisken patents was that the rope was made to travel at a high speed. The rope passed through an arrangement of pulleys and gearing on the plough, which effectively geared down the speed of the rope and converted it into a pull on the plough. The plough itself, in later versions of the Fisken apparatus, was a form of balance plough, which preceded that of John Fowler. Fisken ploughing systems were later marketed by the Ravensthorne Engineering Co., and the patents were extended to allow a further opportunity to develop sales. Although the system had some advantages for smaller acreages, it never achieved really widespread commercial success. The reliability of the equipment was considered to be suspect, probably because of problems of dealing with the speed of the rope with poorly lubricated pulleys.

John Fowler

During the early 1850s there were several other people working with varying degrees of success on the development of systems of cable ploughing. Outstanding among all the pioneers of steam cultivation in terms of success was a man from Wiltshire, John

Fowler, whose ploughing equipment was to dominate the market in Britain, and in some countries overseas, until tractors made steam obsolete on the farm.

John Fowler's career started in Wiltshire where his first job was with a grain merchant. This work failed to satisfy his interest in engineering and he left to join a Yorkshire firm of steam and constructional engineers in 1847, when he was twenty-one years old. It is said that his interest in agricultural mechanization started in Ireland, where Fowler saw the aftermath of the famine caused by the failure of the Irish potato crop. According to the story, he became concerned to improve land drainage and cultivation methods in order to improve the efficiency of food production.

His first interest was in improving land drainage, and he based his development on the existing ideas for mole drains. Although it appears that his work on drainage equipment did not start until 1849, he took out patents on his ideas the following year and he also exhibited his equipment at the Royal Show, which was held in Exeter in 1850, and was awarded a medal by the judges. His success was the result of linking the idea of mole drainage – simply drawing a bullet-shaped object through the soil to make a drainage tunnel beneath the surface – with the more advanced technique of tile drainage. He arranged to draw a series of tiles below the soil, following the metal 'mole', to make a more durable lining to the drainage channel.

At first the power needed to draw the metal 'mole' and the tiles through the soil, was provided by human or animal energy, applied through a windlass. By 1854 Fowler had been able to use a steam-engine to do the job through a cable system, and he was also interested in applying the same idea to cultivations.

Fowler's approach to developing drainage equipment had been to make use of existing theories and equipment, and to adapt and improve them according to his own ideas. He used a similar approach to his work on steam ploughing and cultivating equipment, and in fact throughout his short career there is little evidence that he developed any completely new and fundamental principle. His greatness was in his ability to see how to improve, develop and combine ideas to produce equipment which was better than anyone had made before. He also had an extraordinary ability to gain the confidence and enthusiasm of others, and to build up a commercial organization which set a most impressive standard of teamwork at public demonstrations.

One of the significant milestones in the progress of Fowler's development work was the balance plough, with two sets of ploughshares arranged for two-way working. The balance plough could be worked to and fro across a field, ploughing with alternate sets of shares but always turning the furrows in the same direction. The first Fowler balance plough was built for him by Ransomes, and Sims of Ipswich, and was tested successfully on land at Nacton, Suffolk, in 1856, using a Ransomes portable engine for the purpose of the trials.

During the early years of his steam-ploughing development, John Fowler used a single engine plus an anchor to hold the rope at the opposite headland. In 1856, two years after the idea had been demonstrated by Lord Willoughby d'Eresby and ten years later than John Osborn of Demerara, John Fowler began using two engines, one at each side of the field and linked by the cable. However, it was with a single-engine set that he competed in the steam-ploughing trials at the 1858 Royal Show at Chester and won the Royal Agricultural Society's £500 award for the first steam-cultivating equipment to meet the Society's high standard of work and economy. This award had been withheld for several years because the required standards had not then been achieved, and Fowler's success earned great prestige as well as cash.

The achievement at Chester was one of a remarkable series of awards won by Fowler equipment. The company's skilled and enthusiastic demonstration teams were to earn distinction throughout Britain and in many countries overseas, playing a major part in the commercial success of the company.

This commercial success was becoming apparent in the late 1850s. Various arrangements had been made by John Fowler to have his equipment manufactured for him by other companies, but by 1860 the volume of business had increased sufficiently to justify

Some Savage ploughing engines had the cable drums incorporated in the driving wheels, so that only one power transmission system was required.

his own factory. This was established in partnership with William Hewitson, who had considerable experience of engineering production, at Hunslet, Yorkshire. John Fowler and Company, as the business was called, developed rapidly, with sales concentrated mainly on double-engine ploughing sets. Overseas business was a major factor in the company's prosperity, and for many years the Fowler name was pre-eminent in steam cultivation throughout much of northern and central Europe, and in other countries further afield.

John Fowler saw little of the success which he had established.

On medical advice he had to cut down on the amount of work he was doing, and he moved some distance away from the factory. This was in 1864, and later in the same year, when he was only thirty-eight John Fowler died.

British developments in steam ploughing after the establishment of John Fowler's success, were dominated by his company, and the continuing contributions of other individuals and companies tended to be over-shadowed.

One of the most distinctive types of ploughing engine to be sold commercially was that developed by the Gloucester firm of W. Savory and Son. These were first shown in 1861 and were manufactured in small numbers for several years. The cable winding drum used by Savory was in the form of a huge cylinder, completely encircling the boiler. This may have given good torque characteristics and a simplified drive to the drum, but would have been more costly and complex than the horizontal drum favoured by Fowler.

Chandler and Oliver devised an ingenious means of driving winding drums for a ploughing cable, mounted on the back axle of a portable engine. The drums were located on either side of the firebox, and between the box and the wheels. This arrangement was suitable only for the roundabout system, as with a two-engine system, for example, it would have been necessary to move the engines sideways to keep the plough moving on to uncultivated ground. This idea appeared briefly in 1856.

An alternative was to use the driving wheels of a self-propelled engine as the winding drums. Savage of King's Lynn, Norfolk, used this idea in 1878. Both driving wheels had a deep groove around the circumference. When the engine was travelling the groove was covered for protection. To operate the drums, the engine was jacked up so that the wheels would turn freely, and the covers were removed to expose the cable wound round in the grooves. This system allowed the drive mechanism to the wheels to power the cable drums as well as to propel the engine.

6

FARMING WITH STEAM

One of the less obvious ways to help justify the cost of a steam-engine on a modest acreage, was to use some of the steam for 'cooking' animal feeds. The idea appears to have become quite widespread in Victorian England, and was often the subject for favourable comment in the farming press. The comment was based more on farmers' observations than on scientific evidence, and some of the claims made for the advantages of feeding hay and straw 'steamed' were probably optimistic.

An early report on the subject was published in the *Farmer's Magazine* in 1844, based on a talk given to members of the Burton on Trent Farmers' Club in Staffordshire. The talk was given by Mr John Lathbury, a local farmer and owner of a stationary engine purchased for threshing and to operate a chaff cutter and a mill. The published report suggests that Mr Lathbury was not particularly convincing when describing the advantages of using the steam-engine for working his barn equipment. One of the economies which he stressed as being a result of installing the engine, was that steam threshing was less arduous for his men than when threshing with flails, and because of this, Mr Lathbury was able to reduce the amount of beer bought for the threshing gang.

Mr Lathbury does not appear to have had any doubts about the secondary use for his engine, which was to provide the steam to 'cook' feed for his yarded cattle. In his experience, the effect of steaming was to improve both the digestibility and the palatability of bulky foods such as hay and straw.

The Burton on Trent farmers heard that steaming had much the same effect as fine chopping, and that when hay was steamed its feed value was restored almost to the level of fresh green grass. This had been proved on Mr Lathbury's farm, at least to his own satisfaction, in a feeding test he had carried out on some beef cattle.

When the feed was steamed, he had been able to include a greater proportion of cheap straw and to economize on the quantity of hay. The difference was considerable. In the unsteamed ration he had included 20 tons of straw and 90 tons of hay. In the steamed ration the amount of straw was increased to 60 tons and the hay was cut to 50 tons.

As the nineteenth century progressed, interest in the benefits of steaming feeds grew, and there was a wide selection of steam-raising equipment on the market for livestock farmers who did not possess an engine. A report in the *Royal Agricultural Society Journal* of 1892 made steamed cattle food sound almost like cordon bleu cooking. With steamed straw chaff, said the article, there was a 'grateful aroma, delicate flavour and nice, palatable condition'. Another reference in the same volume of the *Journal* suggested that by steaming, even hay which had been spoiled by weather, could be made perfectly sweet and wholesome again. 'Even the worst, although absolutely white with must (mould) has been known after being chaffed and steamed, to be readily devoured by all kinds of stock.'

Subjective judgements such as these on the value of steaming livestock rations, encouraged many farmers to test the idea for themselves. But on other aspects of using steam power on the farm there was ample information available based on thorough and careful experiments.

Facts and figures on almost any aspect of using steam on the farm was available in most of the more advanced agricultural countries, and this was especially true in Britain during the second half of the nineteenth century. An increasing fund of knowledge was built up, from experiments and from commercial farming, and this information was readily available to anyone who cared to read or to listen. The farming press of the day carried columns of comment, news, pictures and lengthy feature articles, catering for the widespread interest in steam power on the land. Farmers' organizations throughout the land arranged talks and discussions for their members to highlight developments in the application of steam. Steam-engines and equipment were popular attractions at

The steering mechanism for self-propelled engines was developed after experimenting with alternatives. Several manufacturers relied on a horse for steering; Thomas Aveling from Kent provided a fifth wheel with tiller control, and there were various arrangements with a steering wheel located at the front of the engine, as on this Clayton design.

(See facing page.)

agricultural shows, ploughing matches and demonstrations. Efforts were made to inform the general public of the changes taking place on the farm. Agricultural steam-engines were featured at the Great Exhibition of 1851 in London, and there were well-informed articles about steam cultivation in publications such as the *Illustrated London News* and *The Times*.

Those who helped to gather the facts and actively encouraged progress in making use of steam power, include organizations such as the Farmers' Club in London, and various land-owning members of the British aristocracy. Much the greatest contribution came from the major agricultural societies, and especially from the Royal Agricultural Society of England. The work of the R.A.S.E. in encouraging progress, both in the design of equipment and in the efficient use of steam power on the farm, was of immense value. Farming benefited and so did the machinery manufacturers, and the R.A.S.E. played a significant role in maintaining British leadership in applying steam to agriculture throughout most of the nineteenth century.

The R.*A.S.E. Journal*, published annually, was the outstanding source of facts and figures on almost any aspect of farming with steam. Much of this information came from the Society's own trials and investigations. These included carefully measured performance comparisons between different makes of engine or equipment, and some extraordinarily comprehensive and detailed investigations of results achieved on the farm. Some of these surveys, conducted on a national scale, provided a mass of data which must have been of great value, both to the farmer considering a new steam-engine or ploughing system, and to the machinery manufacturer anxious to improve his equipment.

Engines and other equipment were tested annually by the Society in a series of trials which travelled the country with the great agricultural shows. The trials were conducted with great attention to detail and were supervised and judged by experts. To ensure uniformity, both between engines and from year to year, every portable or stationary engine entered for these trials had to use coal supplied by the Society. This was of a uniform grade purchased year after year from the same coal-mine in South Wales, and the chemical analysis of the coal was published to make the performance figures more valuable.

The performance figures published for engines after the annual trials are an indication of the rapid improvement in design achieved by manufacturers about the middle of the last century.

Year	Location of Trials	Performance of winning portable engine in lbs of coal burnt per h.p./hr
1849	Norwich	11·50
1850	Exeter	7·56
1852	Lewes	4·66
1853	Gloucester	4·32
1854	Lincoln	4·55
1855	Carlisle	3·70

(Table based on figures published in *The Engineer*, 7.11.56.)

The various trials and competitions organized by the R.A.S.E. achieved a high standard of thoroughness and objectivity, and manufacturers coveted the publicity value of an R.A.S.E. award much more than they valued the cash prize which accompanied it. Not that the task of the judges at these events was always easy. In 1857, when the R.A.S.E show was at Salisbury, Wilts, the four entries in the steam-ploughing competition experienced great difficulty taking part. Mr Collinson Hall's engine miscalculated a turn on the way to the field and missed most of the trial; Mr

Fowler's engine was also late arriving but then worked well; Mr Boydell's engine was the only one to arrive on time, but it failed to plough properly. The fourth entry was a ploughing system developed by Mr J. A. Williams, a farmer from Baydon, Wiltshire.

Something went badly wrong with Mr Williams and his equipment, as the official report of the judges indicates: 'The judges regret to be compelled to add that the extreme discourtesy of his language and conduct towards themselves rendered their duties in the inspection of his work painful and unpleasant in a manner they never before had occasion to experience at the meetings of the Society.'

The matter was not allowed to rest with a public reprimand for Mr Williams, however. He and the other three entrants got together as signatories to a letter which was widely published in the farming press. This criticized the Society for its handling of the steam plough event, and declared that the circumstances of the trial could only have damaged the cause of steam cultivation. An editorial item in the *Farmer's Magazine* which carried this letter, had criticism for all concerned. It described the trial as a 'mockery', and it also criticized the grammar and style of the letter from the four competitors.

Usually the Society's events were well organized, and these together with comprehensive reports from commercial farms produced quantities of performance figures and costings. At the Canterbury Show in 1860, only three years after the fiasco at Salisbury, steam ploughs showed their pace producing outputs which farmers could use as a yardstick. The highest output came from a Fowler 12 h.p. engine, manufactured by Kitson and Hewitson, with a windlass and rope system. The Fowler demonstration team ploughed at the rate of six acres in ten hours, and this was calculated to cost 8s. 4d. (42p) an acre. A Robey engine with a modified form of Smith's roundabout system achieved 3·5 acres a day at an estimated cost of 11s. 8d. (58p) an acre. The cost for the Fowler system was said to represent a saving of 60 per cent compared to the cost of ploughing with horses, and the Robey costing was half that of horse work.

A different approach to assessing the costs of steam ploughing appeared in the *Journal* ten years later, in a report by the Chief Agent to the Duke of Northumberland. He quoted the actual costs of purchasing and operating a set of steam-ploughing tackle, which the Duke had purchased to rent out on a non-profit basis to tenants on his Alnwick Castle Estate.

It was considered that the cost of steam tackle would be too great for most farmers to find, and the investment by the Duke was aimed at providing his tenants with the opportunity to benefit by steam ploughing without the capital cost. The Duke purchased the following Fowler equipment:

2 ploughing engines of 12 h.p.	£1,200
4-furrow plough	£92
7-tine cultivator	£65
800 yards of steel rope	£84
10 rope porters	£10
Large harrow	£50
Water cart with pump	£25

Annual operating costs were calculated at £494, including £75 for interest on £1,500 at 5 per cent, £100 reserve for depreciation, £160 for wear and tear and oil, and £159 for the wages of three men and two boys. To cover these costs the tenants, most of whom farmed 300 to 500 acres, agreed to pay a set scale of charges for cultivations performed by the tackle. These charges included 10s. (50p) an acre for ploughing, 5s. (25p) for the first cultivation and 2s. 6d. (12·5p) for each subsequent pass, and just 1s. 6d. (7·5p) for harrowing. It was expected that this would raise all but about £50 of the annual cost, and the deficit would be met by the Duke in payment for general estate work, including the timber hauling, which the engines would perform. On the basis of the first year of operating the scheme, the Duke's agent rated it a success which might be copied elsewhere. It was estimated that the sale of plough horses plus the annual cost of their keep, then being saved on the tenants' farms, was worth more than £750, and the steam plough achieved better results in the field.

One cost item which some farmers found unexpectedly high in the early days of steam ploughing, was the wear and tear on the wire cables which pulled the plough. This was an expense which varied according to soil conditions, and also depended partly on the efficiency with which the rope was held clear of the soil.

Experiences with Fowler ploughing tackle on an Oxfordshire farm were reported by the farmer, Mr W. J. Moscrop, in an essay published in the *R.A.S.E. Journal* in 1863. On his heavy clay land the cost of wire rope replacement had averaged 2s. (10p) for each acre ploughed. Mr Moscrop believed this figure was above average because he could not plough his land with more than two furrows, and the rope therefore did more work per acre than on land where three or more furrows could be used.

In the same issue of the *Journal*, Algernon Clark, a prolific writer on steam cultivation, reported the results of a survey he had made of published data on ploughing systems. The cost of rope replacement in his survey, 'Five Years' Progress of Steam Cultivation', averaged 1s. 6d. an acre (7·5p), and in one example, where a Smith roundabout system was used, the wire rope had cost as much as 3s. (15p) an acre.

Generally the 'Five Years' Progress' survey produced a favourable indication of commercial results with steam tackle which was still in its early stages of development. Clark was immensely enthusiastic about the benefits to the soil achieved by steam tillage, and produced examples of heavy clay soils allegedly brought into market garden condition by several years with steam. He estimated that the annual cost of keeping a plough horse was £45 a year – rather higher than most other contemporary estimates. He charged the annual cost of a working ox at £20. As he claimed that on some farms a steam-engine and plough could do the work of thirty to forty plough horses, he found it relatively easy to show steam in a profitable light, but his conclusions probably overrate the cash advantages of ploughing with steam.

Some commentators, although favouring steam power, managed

to mellow their enthusiasm with a degree of caution. A Bedford-shire farmer, Mr R. Valentine, writing in the 1862 edition of the *Journal* quoted his own experiences over several years of threshing with a steam-engine. He was writing mainly to compare the cost and performance of hiring a contractor with a portable engine to come to the farm, or purchasing a stationary engine to thresh the crop.

His report included some less favourable details which others might overlook. When listing the expenses involved in installing a stationary engine, he included the cost of a special below-ground water tank. This was a 9-ft deep, 9-ft diameter brick-lined container which stored water drained to it from rain on nearby roofs, provid-ing a supply of 'soft' water for the engine. This cost £6, and the chimney for the engine house, 40 ft high and made of 7,000 bricks, cost another £20.

The potential output of the threshing drum with the contractor's portable engine was 40 quarters of grain threshed in a day, and this was achieved several times. But Mr Valentine advised farmers not to count on such a throughput, because stoppages and hindrances and short working days brought the true average on his farm over several seasons, to only 20 quarters a day.

Labour costs for contract threshing were surprisingly high in Mr Valentine's experience. He had to find board, lodging and beer for the engineer and the feeder, who came with the engine. In addition the farmer must find a full team of ten men and three boys, making a gang of fifteen with the contractor's men. This meant a total cost in wages, beer and board and lodging, amounting to £1 10*s*. (£1. 50p) a day.

Mr Valentine warned that any attempt to reduce the size of the team would mean lowering the efficiency of the complete job, and he thought that in many instances there would be even more people employed in the team. He did suggest, however, that three men could be saved if an elevator was used to stack the threshed straw, instead of forking it manually.

The labour involved in forking straw from a threshing machine to a stack, created problems especially in the grain-growing areas of the United States and Canada. The answer there was to have a pneumatic conveyor, powered from the steam-engine, to blow the straw away from the drum. The conveyors, known in America as wind stackers, saved a substantial amount of manual labour, and with steam-threshing outfits already numbered in thousands by about 1890, the stackers clearly had a good commercial future.

A group of shrewd businessmen in Indiana recognized the growing market for this equipment, and they also discovered that it was possible to buy the key patents to the wind stacker. They formed a company named the Indiana Manufacturing Company, which purchased the patents in 1891. In spite of the company name, the partners did no manufacturing at that time, but simply held the patents and charged a royalty on every wind stacker sent out by the threshing-machine manufacturers. Every stacker on which the royalty had been paid carried a 'Happy Farmer' medallion showing the slogan 'The Farmers Friend Stacker'. The royalty fee was $30 on every stacker sold, which brought in a very large and almost entirely profitable income to the partners while the patents lasted. In 1901 the number of stackers bearing the 'Happy Farmer' sign sold in that year, was approximately 9,000.

In his book, *Meanwhile . . . Down on the Farm* (published by New-Wood Press, Woodburn, Oregon) author Allan Herman reproduces one of the advertisements run by the Indiana Manufacturing Co. in the late 1890s, to promote a wind stacker competition. $500 in gold was offered in prizes for the best straw stacks built using a wind stacker only, and no manual labour. The entrant had to provide photographs of the stack, supported by signed statements declaring that the stack was built only by means of the stacker. The aim of the competitions, which returned a little of the royalty money to the farms and custom threshing crews, was to encourage better stack building in order to advertise and sell more patented wind stackers.

The wind stackers, with their pneumatic operation, absorbed a good deal of power, and these together with the demand for greater output from the threshers, encouraged manufacturers to offer more powerful engines. It was in adapting steam power to mechanize the harvesting of crops that American manufacturers were most successful against Britain's old-established leadership in farming with steam. American companies produced big capacity threshers and included such advanced features as automatic bagging attachments and self-feeders as well as the wind stackers. In California the Holt and Best companies produced combine-harvesters which were pulled by steam-engines or by large teams of horses or mules. The harvesters had wide, offset cutter bars with beaters taking the crop to what was in effect a moving threshing machine. According to Allan Herman, one of the Best combines with a traction engine pulling it, was claimed to harvest up to 100 acres a day on the big, flat California fields. These machines were in use in the mid-1890s, and some of the biggest machines took a 40-ft swath. The combine-harvester was sometimes powered by steam, as well as pulled by the traction engine. A pipe from the traction engine boiler took steam to the auxiliary engine on the harvester.

One of the first British attempts to harvest with steam power was a public demonstration organized by the Royal Agricultural Society of an Aveling and Porter traction engine linked to a Cross-kill reaper. The demonstration was part of the trials of various makes of reaper held in August 1876 on the Earl of Warwick's estate near the site of the present R.A.S.E. showground, near Leamington Spa, Warwickshire.

According to a report of the trials published by *The Implement Manufacturers' Review*, the steam-powered reaper attracted much interest, and plenty of initial scepticism. The reaper was pushed by the traction engine, and was powered by a chain drive from the engine flywheel hub. In order to keep the complete unit manoeuvrable, the designers had fitted a gantry and steam-operated lifting mechanism to the front of the engine. By this means the reaper could be lifted 6 ft above the ground so that corners could be negotiated more readily. The reaper had a 6 ft cut, and the output was estimated at 30 acres a day cut for the cost of only 8 cwt of coal. Cost of the complete unit was £560.

The Implement Manufacturers' Review report stated that this was the only unit of its kind then working in Britain. Steam harvesting had also been attempted in the United States before 1876, as well as in Britain where Richard Hornsby and the Beverley Iron Co. had tested experimental designs.

Straw-burning steam-engines, especially for threshing, were of great value in some of the grain-producing areas of eastern Europe and North and South America where coal was expensive to transport. John Head, then a partner in Ransomes of Ipswich, and a Russian engineer by the name of Schemioth, designed the first practical straw-burning conversion in Europe in the early 1870s. This development earned Ransomes valuable export business after the kit was shown in Vienna in 1873. Reynold Wik, in his book *Steam Power on the American Farm* quotes contemporary reports showing that straw-burning engines for threshing were developed in California in 1863, and the idea was widely accepted there three years later, indicating that in this development America had lead Britain.

With vast acreages to plough, and with a shortage of manpower to do it, American and Canadian farmers in the great arable farming areas demanded increasing pulling power from their steam-engines. Some of the immense engines produced from about 1900, to meet this demand, have survived in North American museums. One of the most popular of these engines was the 110-h.p. Case, several of which still exist, including one at the Manitoba Agricultural Museum with power steering. The pride of the Western Development Museum collection at Saskatoon, Saskatchewan is a 120-h.p. Reeves engine, which the Museum sometimes demonstrates pulling a 20 bottom, or furrow, plough at 2½ m.p.h.

Some of these big ploughing engines took part in the annual competitions for tractors and steam-engines which were held from 1908 in conjunction with the Winnipeg Industrial Exhibition. This series of trials started as a public demonstration and comparison of the gasoline tractors which were then coming into Canada in large

84

numbers, sometimes poorly designed and with inadequate service and spares back-up. From 1909 until the series of competitions ended in 1914 the scope of the competitions was expanded to include classes for steam-engines, and the detailed results sheets are an interesting comparison between the performance of some of the finest steam-engines built, and the tractors which were challenging them.

The Case 110-h.p. steam-engine competed with considerable success at Winnipeg. This model was entered in 1909, 1910 and 1912, and in each of these years it won its class against up to four competitors.

In the 1910 results sheet the big Case, with its single 'square' cylinder of 12 in. diameter and 12 in. stroke, registered 129 h.p. in the half-hour maximum brake test. During this test the engine used only 278 lb of coal, compared with 517 lb used by the 90-h.p. Avery and 511 lb by the Rumely 120 h.p. The h.p. hours per 100 lb of fuel averaged 23·4 for the Case, 12 for the Avery and 17·6 for the Rumely. In the ploughing test the Case pulled 12 bottoms or furrows of 14-in. width, exerting almost 11,000 lb of drawbar power, and ploughing an average of 3·99 acres an hour for more than 8 hours. The efficiency of the Case design showed up in the fuel consumption results for the ploughing test, with an average of 99 lb of coal used per acre ploughed, against 120 for the Rumely and almost 150 for the Avery. The best ploughing output in that year by a gasoline tractor was 2·54 acres an hour for the International Harvester 45–55. The tractor weighed approximately half as much as the 40,000 lb of the Case steamer, and the list price of the tractor at $2·700, was almost $1,000 less than the steamer. The smallest steam-engine entered in 1910, the Case 36 h.p., had ploughed 1·31 acres an hour, weighed 17,475 lb and cost $1,812 and 50 cents.

The Engine Driver

Among those who benefited from the use of steam on the farm were the agricultural labourers who worked with the engines. In

Victorian Britain the payment of a premium wage to the 'engineman' and sometimes to the steam ploughman as well, seems to have been a general practice. At a time when agricultural rates of pay were far from generous, the small amount of extra money would have been welcome.

In the costings for the Duke of Northumberland's steam-ploughing investment, already referred to, the foreman engineman was to be paid 25*s*. (£1·25p) a week, while the ploughman would get 18*s*. (90p) and the boys, who hauled water and coal and were to help with odd jobs, got as little as 6*s*. (30p) a week. This was in 1870. In 1863, Mr W. J. Moscrop, whose report on his own experiences with steam ploughing in Oxfordshire has already been referred to, devised an incentive scheme for his men based on the number of acres ploughed. One of the biggest factors limiting performance, he found, was the time lost through breakdowns – often a result of operator carelessness. The bonus payments, additional to the basic wage, were paid to all the men involved in the ploughing, and proved to be an effective means of improving standards of maintenance and care. A year earlier, Mr Valentine, whose experiences with steam threshing have also been referred to, paid his 'engineer' 6*d*. a day (2·5p) extra whenever he was actually using the engine. This modest addition to the man's wages was quite sufficient, said Mr Valentine, to induce the man to keep the engine clean and in good order, and to stay on an extra half an hour or so in order to do so. Mr Valentine knew that some other farmers paid the extra throughout the year, whether the engine was in use or not, but this was a waste of money, he thought.

A Lincolnshire contractor, quoted in the immense 'Report on Steam Cultivation' in the 1867 R.*A.S.E. Journal*, employed his engineman and ploughman on a 10-month contract each year. He paid the two men 20*s*. (£1) a week each, plus overtime. The over-time payment seems to have been popular, for in some instances the men had operated the ploughing tackle from 3 a.m. through to 10 p.m., although they normally worked a 10-hour day. When bad weather halted ploughing for up to a week, the men would be paid the basic rate and were expected to busy themselves with maintenance work. In the slack season from the beginning of June to the end of July, the men were paid 1s. (5p) a day, and could do what they liked with the time. Usually they could find casual work on farms in the area, and this also applied from Christmas for two months, when the men were not under contract. The man who operated this particular scheme, and applied it to the men working his two sets of cable ploughing tackle, was Mr J. Smith of Louth, who had previously worked as manager of the Fowler demonstration and show team, but had since set himself up as a ploughing contractor.

In the same *R.A.S.E. Journal* another incentive scheme was described for a contracting business in Yorkshire. The company, based at Wakefield and operating two sets of tackle was the West Riding Steam Ploughing, Cultivating and Thrashing Co. Ltd, one of the oldest concerns of its type in Britain, founded in 1862. The foreman for each of the sets of ploughing tackle earned a basic wage of 15s. (75p) a week, and received also 10 per cent of the gross earnings of his equipment after paying the wages of the men in his team.

Another scheme to encourage higher standards of work by enginemen was introduced in 1864 by the Peterborough Agricultural Society. This was an annual award of £10 paid by the Society to the top entrant in their annual test for operators of portable engines. The scheme was described as a success in a write-up in the 1868 *Royal Agricultural Society Journal*, and was said to have raised the standard of operation and maintenance of engines in the area.

Entry qualifications included at least eighteen months' steam-engine experience, an accident free record, and the usual Victorian stipulations about a sober character. Each candidate would be confronted with two engines, one of which was deliberately out of adjustment. An examiner noted the men's comments on the condition of the engines, and awarded marks accordingly. The examiner was expected to make full allowance for the fact that enginemen were not accustomed to public speaking. The second part of the test was a visit to the farm where the candidate worked, where 'his' engine would be thoroughly inspected by a senior engineer from the Great Eastern Railway.

Men who worked with ploughing tackle, especially when employed on contract work, spent much of their time away from home, and living vans were available to provide a place for them to sleep and eat. With its usual thorough approach to checking up on anything connected with farming, the R.A.S.E. organized a trial of 'Vans for Men Engaged in Steam Cultivation' in 1870. This attracted three entries, each of which was subjected to thorough appraisal.

The first test was on the ventilation system, and to check on this the judges had all the ventilators closed, and a carefully weighed quantity of tallow and other food waste burned on the cooking fire of each stove. After the fires had burned out, each van was tested for smell. The strongest 'stench' lingered in the van entered by Faulkes of Bingham, Nottinghamshire. This proved that the van, with ventilators closed, was the most airtight of the three. Surprisingly, it was the effectiveness of the seal on the ventilators which lost points. A near perfect seal meant that conditions within the van would become unhealthy when the occupants shut themselves in with the fire burning on a cold winter's night.

In other tests the drawbar pull to move the vans, their general construction and design, price, the cooker, and other fittings were all compared. The vans all had beds for five or six men, and the Faulkes van provided mattresses and bolsters for each. The interior of the vans was somewhat similar to that of a modern caravan, with some of the bunks converting to seats for daytime use, lockable food cupboards, and tables which could be completely or partially folded away.

The Faulkes entry, the most expensive at £110, came third. In second place was the Aveling and Porter van, which lacked springing and gave the least comfortable ride, and the Fowler van, which boasted a desk for the foreman's paperwork, was the winner.

7

COMPETING WITH THE TRACTOR

Towards the end of the last century, after nearly 100 years of development, the steam-engine was beginning its main period of success in farming. Improvements in engine design and construction had achieved standards of reliability and performance which became increasingly acceptable.

Although steam was never to make a major impact on world food production methods, there were sectors of the farming industry in which it became significant. Steam was important for powering cotton gins in the American south and cane-crushing equipment on the larger sugar estates in the West Indies. Vast areas of American prairie land were first ploughed with steam power, and steam ploughing by cable was popular for many years on the heavier clay lands of Britain. In areas of large-scale cereals production, steam-engines were preferred for threshing, and in some areas steam traction engines were the main method of transporting produce such as wool and grain to the market or railway.

While steam was making the breakthrough to commercial acceptance in agriculture, a rival power source was being developed and improved. This was the internal-combustion engine, which started as a European development, but was first adapted for farm use in the United States.

Production of agricultural steam-engines began to increase significantly from about 1875. The numbers grew rapidly and then declined again, during the next fifty years. The J. I. Case Company of America was the largest manufacturer of steam-engines for agriculture during this period, and their production figures are an indication of the way the industry grew and then diminished.

Production of Case steam-engines started in 1869 and continued until 1924. During this period more than 36,000 engines were manufactured, almost all of which were portables or traction engines.

J. I. Case production of portable and traction engines

1882	506
1892	572
1902	1574
1912	2252
1922	153

Before 1900, production of tractors was on a small scale, and total output from all manufactures was less than 100 units in 1899. By 1910 sales of tractors in the United States had reached about 4,000, and this had risen to more than 200,000 in 1920. Henry Ford's remarkable Fordson tractor achieved a production of more than 100,000 in 1923 and again in 1925.

The Challenge

The petrol- or gasoline-engined tractor made its first appearance in about 1889 with an experimental machine produced by the Charter Gas Engine Company of Chicago, Illinois. This tractor was sufficiently successful to encourage the Charter Co. to produce four or five more, all of which were sold, apparently, in South Dakota.

The Charter tractor consisted of a single-cylinder engine mounted on the frame and wheels of a Rumely traction engine. The idea of using steam-engine components to carry petrol engines was adopted by most of the tractor designers working on prototypes in the 1890s. The result was that the first tractors were heavy and difficult to manoeuvre, and most were used for stationary work, such as threshing. Some of the early tractors resembled traction engines in appearance, and this was to persist for several years in Britain. Fowlers announced a tractor for cable ploughing in 1912, and a Royal Agricultural Society report criticized its design, with a dummy chimney and boiler, for being based too closely on steam-engines. The Walsh and Clark Victoria ploughing engine, manufactured at Guiseley, Yorkshire, from about 1914, was also based firmly on the steam-engines it was intended to replace. The Victoria had a tall chimney, and the horizontal cable drum was carried beneath the 'boiler', which was an enormous fuel tank with capacity for a week's work between refills.

As long as tractor designers simply copied the steam-engine, in both appearance and function, steam remained competitive. Both machines were heavy and costly, and neither was suitable for the smaller acreage farms which had still to rely on horses. The tractor scored through being ready for work while the steam-engine was building up boiler pressure, but the early tractor was no match for the reliability and simplicity of the steam-engine. Primitive fuel and ignition systems became temperamental in the dust and damp of farmwork, and their complexity frequently baffled the novice drivers who were more at home on a horse or on the footplate of a traction engine.

One of the first designers to break away from steam-engine thinking, and to produce what was then a new type of tractor, was Daniel Albone of Biggleswade, Bedfordshire. His Ivel tractor, produced in 1902, had a 14-h.p. engine on a three-wheel chassis. The complete tractor weighed 30 cwt, and was designed more for field work than for stationary operation. It could pull a three-furrow plough – at least on light or medium soils – and was economical enough to be used for pulling a binder or a grass cutter.

The Ivel tractor, and its more famous and more developed descendants such as the Fordson, Farmall and Ferguson, between them achieved much more impact on the farming industry than 100 years of steam power. Engines rapidly became more reliable and more efficient, and tractor design improved to achieve more output in the field with less weight. Production costs and selling prices were forced down until they came within the range of smaller farms, and at the same time the usefulness of the tractor was increasing with rowcrop designs, rubber tyres and power take-off.

Some steam-engine manufacturers were quick to recognize the challenge of the internal-combustion engine. The Case company in America produced a prototype tractor in 1892 with a horizontal twin-cylinder engine. This experimental tractor proved unreliable because of the ignition system, and Case abandoned the project to concentrate on their steam-engines. The Huber Manufacturing Company of Marion, Ohio also took an early interest in tractors. In 1896 Huber purchased the Van Duzen company, which manufactured gasoline engines, and in 1898 sold thirty Huber tractors, which consisted of the Van Duzen engine on a Huber traction-engine chassis. This was the first make of tractor to sell in such numbers, and Huber remained in the tractor business for more than forty years.

Before 1900 tractors were still a novelty and few saw them as a real threat to the supremacy of horses or steam on the farm. But after the turn of the century the situation changed, especially in North America, where demand for tractors was increasing significantly. Each year more and more companies launched into the expanding tractor market, competing against each other and against the steam-engine manufacturers. As the tractor market developed, there were signs of faltering confidence among the traction-engine companies. Some began to hedge their bets, launching into the tractor market themselves while trying to hold on to their place in the market for steam-engines. By 1914 many of the most prestigious names in the steam-engine industry were competing in the tractor market. These included Avery, Case, Gaar Scott, Geiser, Nichols and Shepard, and Rumely.

Farmers in Britain were not so ready as their American and Canadian counterparts to accept the idea of a tractor, and the U.K. market developed more slowly. The few firms which were making tractors in Britain, had to rely on export sales to keep their factories in production. Steam-engine companies could afford some complacency, right up to the start of the First World War. During the war, a desperate need to increase home-grown food supplies brought in large numbers of American tractors, particularly the then recently developed Fordsons. Before the war only one British

1 Trevithick stationary engine, 1811.

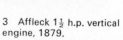

2 Colt's Arms Co. engine, patented 1868.

3 Affleck 1$\frac{1}{2}$ h.p. vertical engine, 1879.

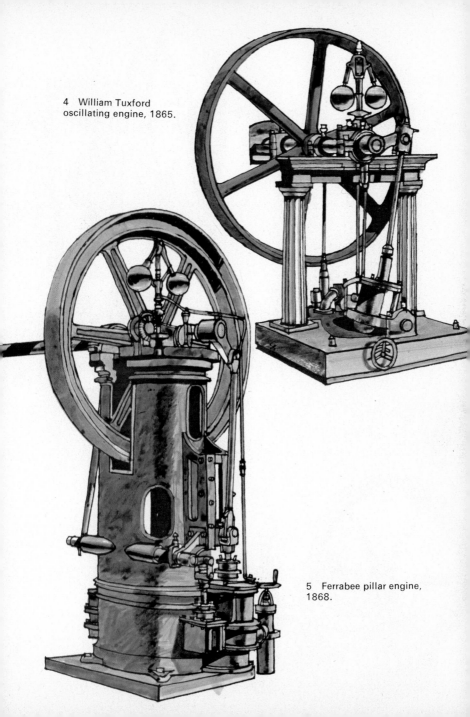

4 William Tuxford
oscillating engine, 1865.

5 Ferrabee pillar engine,
1868.

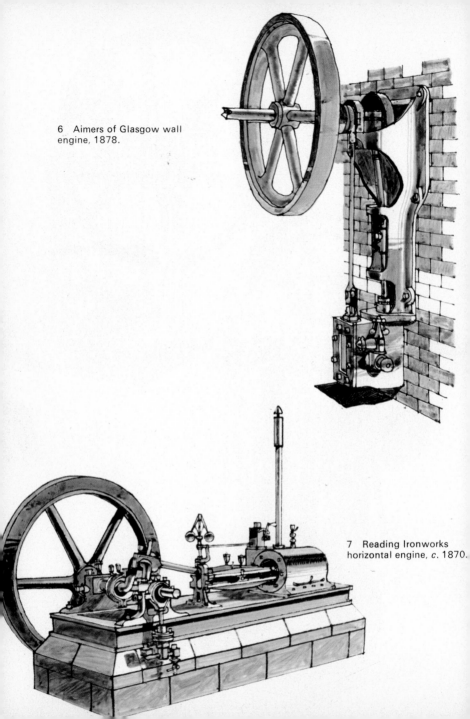

6   Aimers of Glasgow wall
engine, 1878.

7   Reading Ironworks
horizontal engine, c. 1870.

8 Dean portable engine, 1845.

9 Archamboult 'Forty-Niner' portable, made in America, 1849.

10 Smith & Ashby portable farm engine, *c*. 1857.

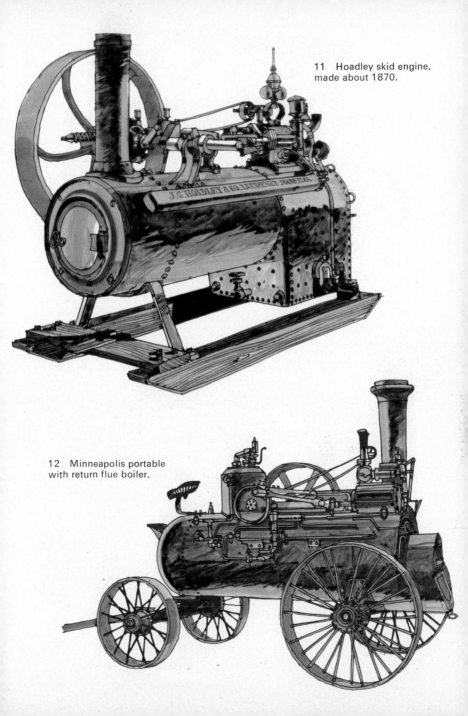

11 Hoadley skid engine, made about 1870.

12 Minneapolis portable with return flue boiler.

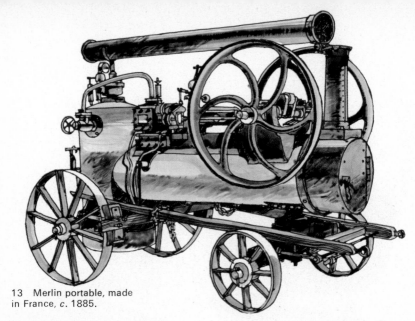

13   Merlin portable, made
in France, *c*. 1885.

14   Wolf engine, Germany, 1862.

15   The first J.I. Case portable engine, 1869.

16   Brown & May 6 n.h.p. portable, 1880.

17 Brown & May portable of 1870, with horse.

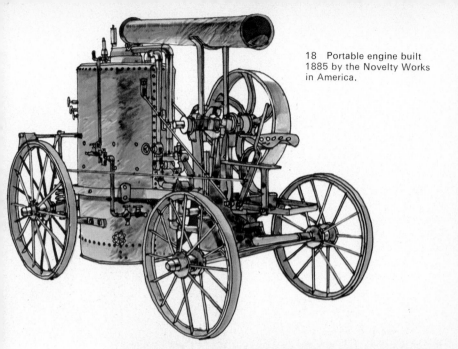

18 Portable engine built
1885 by the Novelty Works
in America.

9 Rustons of Lincoln
rtable engine, 1913.

20 Marshall portable
exported to Ireland in 1916.

21 Ransomes, Sims and
Jefferies portable, 1918.

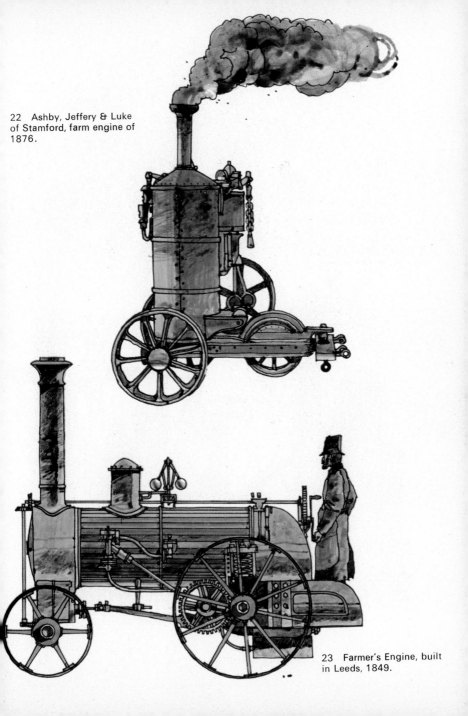

22 Ashby, Jeffery & Luke of Stamford, farm engine of 1876.

23 Farmer's Engine, built in Leeds, 1849.

24   Holcroft screw propelled engine of 1856.

25 Alexander Chaplin & Co. of Glasgow traction engine, 1861.

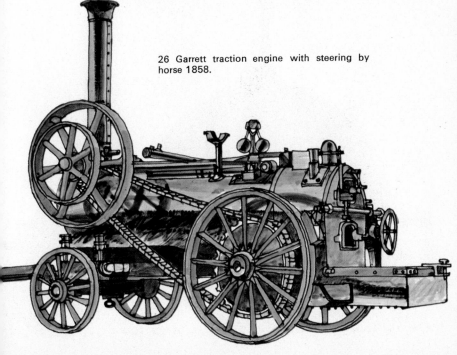

26 Garrett traction engine with steering by horse 1858.

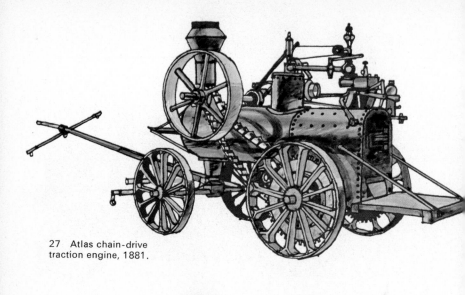

27 Atlas chain-drive
traction engine, 1881.

28 The Blackburn
ploughing engine, 1857.

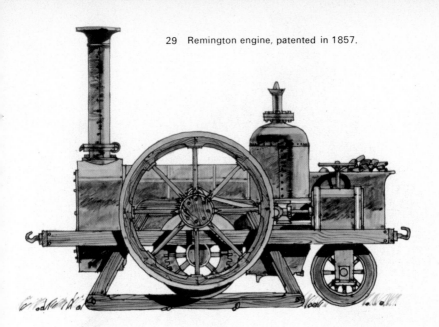

29 Remington engine, patented in 1857.

30 Stratton 'half-track', built in Moscow, Pennsylvania, 1893.

31   June traction engine
with patented safety
chimney, 1899.

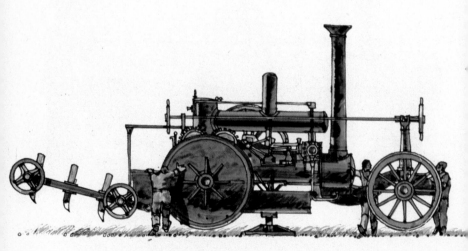

32   Collinson & Charlton ploughing engine, 1857.

33 Case straw-burning traction engine, *c*. 1885.

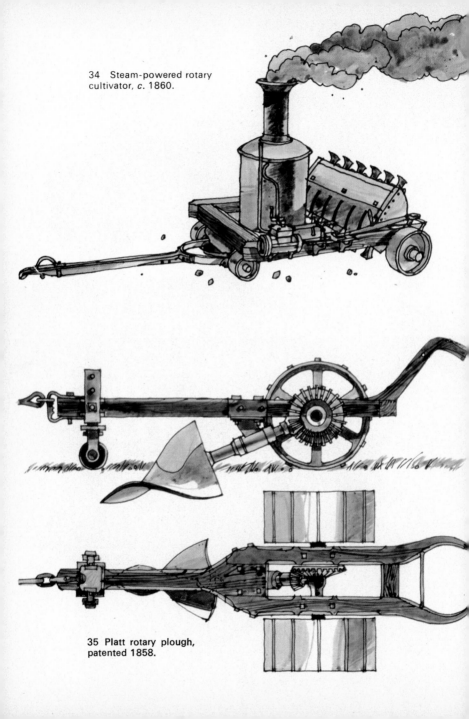

34 Steam-powered rotary cultivator, *c*. 1860.

35 Platt rotary plough, patented 1858.

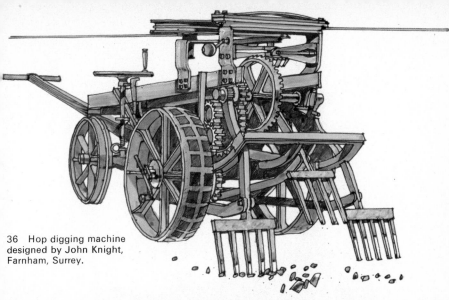

36 Hop digging machine
designed by John Knight,
Farnham, Surrey.

37 Steam-powered rotary plough patented by
James Usher, 1849.

38   Rickett self-propelled rotary cultivator, 1858.

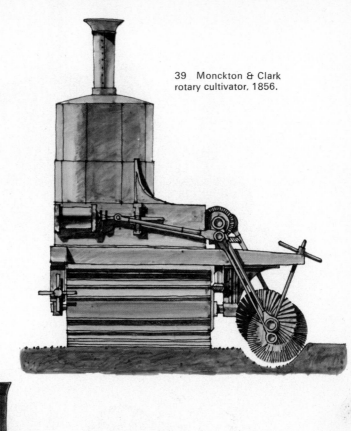

39 Monckton & Clark
rotary cultivator, 1856.

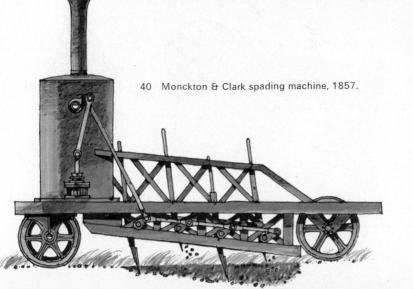

40 Monckton & Clark spading machine, 1857.

41 Grimmer ploughing engine from Lincolnshire, 1881.

42 Frank Proctor's steam-powered rotary digger, 1880s.

43   Savory ploughing engine, 1863.

44   Burrell 'Universal' cable ploughing engine,
advertised 1882.

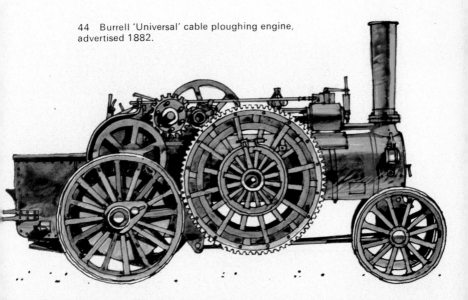

45   Fowler double-engine cable ploughing system,
1921.

46   (*Left* and *Above*) Fowler of Leeds model BB1
ploughing engine, 1921.

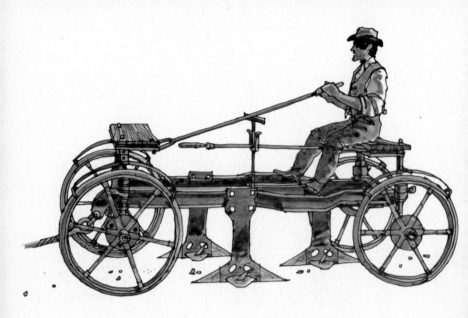

47 Howard of Bedford two-way cultivator.

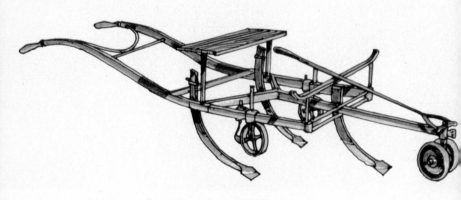

48 William Smith's 1862 cultivator.

49 Sutherland steam plough for moorland reclamation, *c.* 1871.

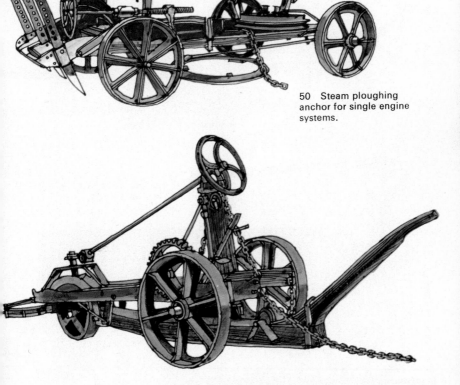

50 Steam ploughing anchor for single engine systems.

51 Knights & Stacey mole plough, *c.* 1885.

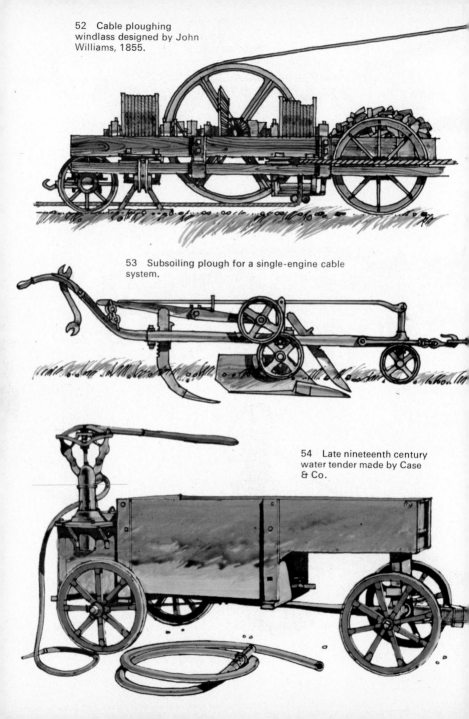

52 Cable ploughing windlass designed by John Williams, 1855.

53 Subsoiling plough for a single-engine cable system.

54 Late nineteenth century water tender made by Case & Co.

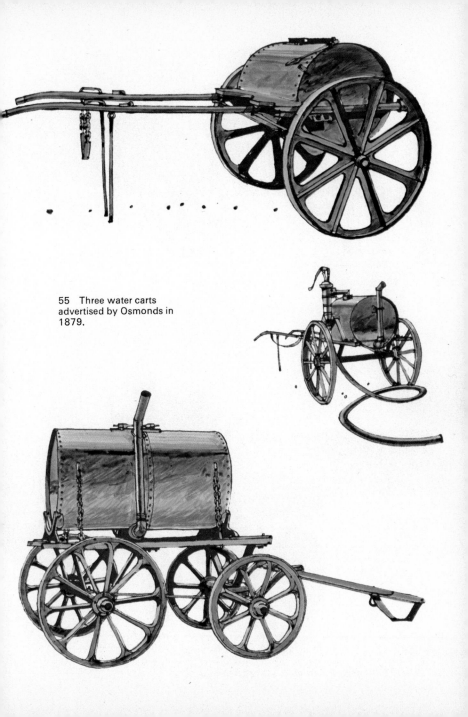

55 Three water carts advertised by Osmonds in 1879.

56 Howard of Bedford straw press, 1880s.

57 Ann Arbor 'Columbia' baler advertised in 1904.

58   Machine for making 'round' bales, designed by
Pilter of Paris.

59   The 'Perpetual' baler, designed in America,
c. 1880.

60   Aveling & Porter traction engine with Crosskill reaper, 1876.

61   Best traction engine
and combine harvester,
c. 1895.

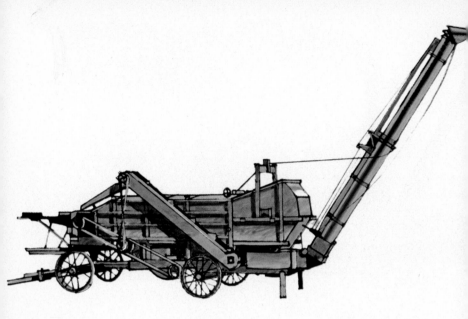

62 Case threshing machine with blower attachment for straw, 1897.

63 Case thresher with mechanical elevator for straw stacking.

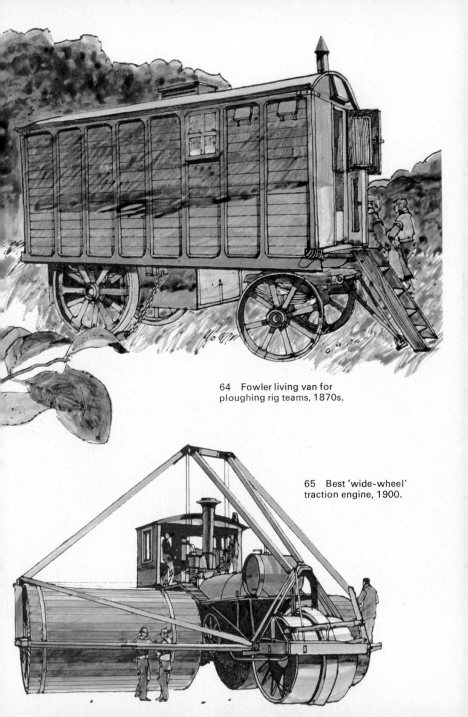

64   Fowler living van for
ploughing rig teams, 1870s.

65   Best 'wide-wheel'
traction engine, 1900.

66   Replica of a Savage traction engine of 1865.

67   Threshing and baling demonstration at a Suffolk
rally, 1976.

69   Ransomes, Sims & Jefferies threshing drum.

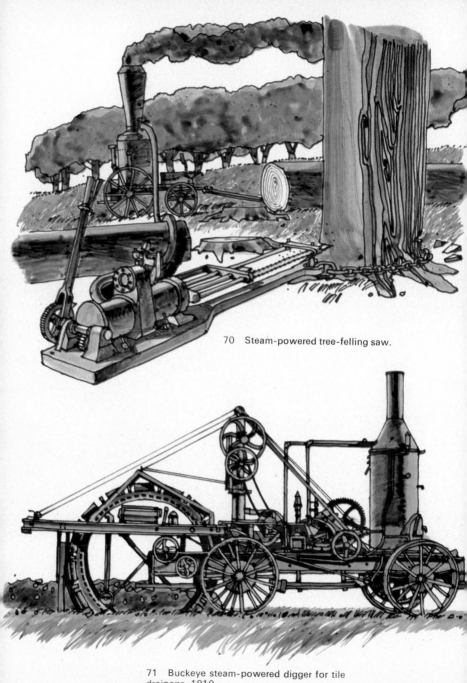

70 Steam-powered tree-felling saw.

71 Buckeye steam-powered digger for tile drainage, 1910.

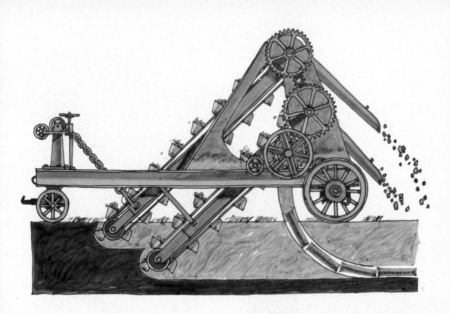

72 Robson & Herdman patent draining machine of 1881.

73 Steam-powered version of the Chicago trench digger, c. 1905.

74 Westinghouse traction engine, made in
Schenectady from 1881.

75   Burrell 8 n.h.p. traction engine, 1876.

76   Aultman & Taylor shaft drive engine, *c.* 1885.

77   Marshall traction engine, 1887, driver's view.

78   Avery 40 h.p. undermounted traction engine.

79   Four-wheel drive Wood, Taber & Morse engine,
c. 1885.

80   Gaar Scott 'Big Forty'
traction engine, 1911.

81   J. I. Case 28-80 traction engine, 1912.

82   Marshall traction engine, 1908.

83   Burrell 6 n.h.p. engine, 1901.

84   Wallis & Steevens traction engine, 1919.

85 Garrett 7 n.h.p. traction engine, 1916.

86 Burrell 'Devonshire' engine, 1909.

87   Allchin general purpose engine, 1911.

88   Clayton & Shuttleworth traction engine, 1923.

89 Ruston Proctor agricultural engine, 1908.

90 Burrell agricultural engine, built 1907.

91   Burrell traction engine no. 3923, built 1922.

92   Fowler of Leeds $3\frac{1}{2}$ n.h.p. engine, 1921.

93   Marshall general purpose traction engine, built
1919.

94   Ruston Proctor agricultural engine, 1914.

95 Ransomes 7 n.h.p. traction engine, 1919.

96 Parade of engines at the 1976 rally at Weeting,
Suffolk.

97 International steam
prototype, 1922.

98 Second International
Harvester steam tractor,
1923.

99 Bryan steam tractor, 1923.

99 Bryan steam tractor, 1923.

100 Sentinel Roadless steam tractor, 1924.

101   American Abell 25-ton traction engine, 1911.

102 Garrett 'Suffolk Punch' steam tractor, 1919.

103   The 'Leeside Rover', built in Ireland, 1971.

steam-engine company, Marshall of Gainsborough, had developed a commercially successful tractor. Others, including Ransomes, Fowler and McLaren, produced either prototypes or commercial tractors selling on a very small scale before or during the war.

While the market for traction engines and portables was being challenged by the tractor, the demand for stationary steam-engines for farm use had already been ruined by the internal-combustion engine. Well before 1900, scores of manufacturers, especially in the United States, were producing large numbers of petrol engines for farming. These were compact, light, reasonably reliable and inexpensive to buy. An engine of about 4 h.p. could be bought for as little as $20 in the early 1900s, suitable for operating pumps, feed preparation and dairy equipment, and for mounting as an auxiliary power unit on the horse-drawn balers which were coming on to the market. The market for small stationary steam-engines, which had flourished in America and Britain until about 1890, diminished rapidly. In Britain many of the petrol engines which flooded on to the market, were imported from America, and large numbers of these and their British-made competitors, have survived for restoration.

Traction-engine manufacturers generally reacted in one of three ways to the growing success of the tractor. Some, including many of the most successful companies, accepted the opportunity to join in a rapidly expanding new market. Some companies simply went on producing steam-engines for a declining market, until they eventually went out of business or were taken over by more progressive rivals. Just a few of the traction-engine firms adopted the third, and more difficult alternative, of designing a better traction engine to compete against the tractor.

Steam Tractors

As tractor design progressed, with rising standards of reliability and lower prices, the gap between the tractor and the traction engine was widening. To bridge that gap, with a traction engine which could compete against the tractor, demanded a completely new approach to engine design. Several companies produced new models which showed a great deal of originality, and which were probably great improvements on the traditional design of traction engine. None of these was a lasting commercial success, and all failed to meet the objective of competing commercially against the tractor. Although these were failures, they were the last phase of traction-engine design, and were probably among the best traction engines ever produced.

Probably the last company in America to announce a completely new traction-engine model was the A. D. Baker Co., of Swanton, Ohio. The Baker 22–45 h.p. model was announced in 1925, and described as a 'Steam Tractor'. It weighed 14,300 lb, which was little heavier than many American tractors at that time, developing similar power. The styling was more like that of a tractor, and this was emphasized by a radiator with fan which was mounted ahead of the boiler acting as a condenser to conserve water. Advanced features of this traction engine, apart from the condenser, included a high-pressure (300 lb) boiler, a tandem compound engine, automobile type front axle and an automatic stoker for the fire. Dimensions were kept as compact as possible, with a wheelbase of 8 ft 11 in. and a minimum turning circle of 16 ft.

The Baker Steam Tractor arrived at a time when tractor design was progressing rapidly, and when American farmers had almost stopped buying traction engines. In spite of its advanced specification, the Baker machine attracted little interest or demand. It seems that the Baker company may have had little faith in their challenge to the gasoline tractor, for only a year after announcing the steam tractor, the company announced a new 25–50 h.p. gas tractor, their first departure from steam.

In Britain one of the first of the new generation of traction engines was an unconventional lightweight from the Summerscales company in Keighley, Yorkshire. The Summerscales steam tractor had a vertical boiler mounted centrally on the chassis. The engine had four cylinders in a vee formation with poppet valves operated from a camshaft. Drive from the engine was by Renold-type chain to the rear wheels. There was a single wheel at the front for steering, a feature which was appearing on some British and American tractors at about the same time. The engine was rated at 25 h.p., and the weight in working trim was less than 4 tons. The Summerscales tractor was designed during the war, and made its major public appearance at the 1919 tractor trials at South Carlton, Lincoln. It appears that only two machines were produced.

Another break from tradition produced at about the same time as the Summerscales, was the 'Suffolk Punch' or Agrimotor, designed and produced by Garrett of Leiston, Suffolk. The prototype was tested with a four-furrow plough on Foulness Island off the Essex coast in 1917, in a bid to prove that the cost of ploughing by steam could be competitive with tractor costs.

The layout of the Suffolk Punch was arranged so that the driver was seated right at the front, with excellent forward vision and little visibility to the rear. The locomotive-type boiler was positioned so that the firebox was located behind the driver, and the chimney was at the back. The external appearance of the steam tractor was similar to that of some tractors of the time, with an impression of compact power which made the name, Suffolk Punch, most appropriate.

The technical specification was advanced, and included a superheater to raise the steam temperature to 600 deg. F., for greater economy. The two axles were sprung, with coil springs at the front, and there was Ackerman steering with a lorry-type steering wheel. The double-crank compound engine was rated at 37 to 40 b.h.p., and the engine was designed to operate efficiently on inferior grades of coal.

In spite of its advanced specification and appropriate name, the Suffolk Punch failed as a challenge to the tractor and only eight were built. With the imported Fordson tractor selling for £225 in 1919, the prospects for steam looked gloomy. In spite of this, the Sentinel Company at Shrewsbury, decided in the early 1920s to launch into the traction-engine market for the first time. The Sentinel company had made their reputation with steam vehicles for road transport. Their first steam tractor was actually based on steam waggon components and was designed as a rough terrain transport vehicle.

Having developed a tractor for rough conditions, the next stage of development, appearing in 1924, was a modified version for agricultural and forestry work, including ploughing. This had suitably reduced gear speeds and a choice of either large rear steel wheels or Roadless tracks. The tracks, developed from the First World War experience, helped the performance, but added to the price. The Sentinel–Roadless tractor cost £1,250, but was claimed to exert a 6-ton drawbar pull and to have climbed a one in one gradient on test. An improved later version, designed for operation in Africa, pulled twelve furrows on test, equal to 18 acres a day.

The Sentinel–Roadless was reasonably successful in Africa, and about fifteen were sold. This encouraged Sentinel to introduce a new wheeled version, which was christened the Rhinoceros, presumably with the African market in mind. The 'Rhino' had a 9 ft 9 in. wheelbase, and a locomotive-type boiler designed to work at 275 p.s.i. The tractor was rated at 86 b.h.p. and ambitiously priced at £1,800. The standard version of the Rhinoceros was coal fired, but wood-burning and oil-fired adaptions were available. Maximum drawbar pull was 22,400 lb, and a winch was fitted for timber work. Ploughing output was claimed as 30 acres a day with twelve furrows on light land.

In their book *The Sentinel* (published by David & Charles), the authors W. J. Hughes and Joseph Thomas, estimate that eight of the Rhinoceros steam tractors were sold, all for export. Although the Rhinoceros was outdated by the internal-combustion engine, they claim that it was possibly the finest and most versatile steam tractor ever built. Sentinel persisted with steam vehicles longer than most of their competitors, and the company was eventually taken over by Rolls Royce.

8

A FUTURE FOR STEAM?

A small market for traction engines and portables survived until 1939, when production ceased because of the war. During the war a large number of engines finished up as scrap metal, and others not in use as part of the drive for increased food production, were allowed to decay.

The war really signalled the end of steam power in commercial farming. Immediately after the war there were isolated instances of portable engines being produced in Britain to meet orders from overseas. Some manufacturers survived on a modest scale for a few years making steam-rollers, for which there was a small demand into the early 1950s. But with limited exceptions, the commercial use of steam power on the farm was almost dead. There are still some steam-engines working on farms, both in Britain and in North America, but the reasons for using them are more often nostalgic than economic. There is also the occasional demand for steam power to do a special job, such as dredging a lake with a scoop wound in on the drum of a cable ploughing engine – but this level of demand can hardly provide a regular income to a contractor.

As more and more engines were sold from farms and by contractors during the years following the war, the value of an engine was measured by its scrap metal content. Large numbers of fine engines rusted beyond repair, or ended their days in scrap yards.

Fortunately the steam-engine preservation movement began to develop during the early 1950s, in Britain and America, when there was still a substantial stock of engines to be preserved. Steam power has an immense appeal which other forms of power, such as an electric motor, are unlikely to match. The appeal of steam is sufficient to encourage people to spend considerable amounts of time and money restoring old steam-engines and maintaining them in show condition. The appeal of steam is also strong enough to encourage large crowds to the season of traction-engine rallies, which have become one of the biggest attractions during the summer.

Traction-engine rallies, or threshermens' reunions as they are sometimes called in the United States, are popular in many countries, including Australia and New Zealand, Ireland and some European countries. They range from small, informal affairs, with perhaps half a dozen engines, up to major events attracting engines and spectators from 100 miles away. The more elaborate events usually include vintage tractors and other attractions. One of the biggest annual events in Britain with steam as the main attraction, is the 'Great Working of Steam Engines' organized by the Dorset Steam and Historical Vehicles Benevolent Fund. The main objectives are to provide entertainment and interest to a large number of people taking part or watching, and also to raise large amounts of money for charity. One recent 'Great Working' attracted nearly ninety steam-engines, including rollers and showmens' engines, plus more than 200 tractors and more than 500 stationary engines. Other attractions included displays of country crafts, and there was also an old-fashioned steam fair, vintage aircraft, old cars and lorries and a display of model engineering.

The attractions of owning steam-engines – including their investment value – has meant rising prices and it is now difficult for anyone with modest resources to buy an engine which does not require substantial restoration. An auction sale in Hampshire in 1976, when items from the Wessex Machinery Museum were sold, indicated recent values. A Fowler BB ploughing engine, one of a pair built in 1910, attracted a top bid of £4,500, a Marshall portable engine sold for £950 and a Burrell agricultural traction engine of 1921 attracted a bid of £6,000. Top price for a steam-engine at the sale was the £28,500 bid for a Burrell 7 n.h.p. showman's road locomotive built in 1921.

A number of the more interesting steam-engines have been preserved in museums, and some of the great museums of technology have agricultural engines on display. The Deutsches

Museum in Munich has a display of model Fowler ploughing engines, plus the original Wolf portable engine of 1862. The Smithsonian Institution in Washington has several engines including the first Case portable of 1869 and also a much later Huber return flue traction engine. The Science Museum in London has a particularly interesting collection of early stationary engines, including beam engines of the type which were first appearing on farms around 1800.

A recent development has been the opening of museums specializing in farming and rural life. Many of these in Britain feature a small number of agricultural steam-engines, with the collection at the Bressingham Steam Museum near Diss, Norfolk, being outstanding.

By British and European standards the displays at some of the specialist museums in the United States and Canada are enormous. The prairie provinces in Canada have their own museums, and these included twenty-seven traction engines recently at the Manitoba Agricultural Museum and a similar number at the Reynolds Museum in Alberta. The Greenfield Village and Henry Ford Museum at Dearborn, Michigan, has probably the most comprehensive and interesting collection of agricultural-type steam-engines in the world, with more than sixty engines. These include such rarities as a Howard ploughing engine from England, a Merlin traction engine from Vierzon, France, and an 1865 Cooper skid engine from Mt Vernon, Ohio.

With such widespread interest in steam-engines of various sorts, it is inevitable that there must be a flow of publications to cater for the enthusiast. This includes a particularly wide selection of books dealing with the development of steam ploughing with traction engines in general or covering the history of leading companies in the steam-engine business. There are also journals of various sorts, including *Western Engines* published in Portland, Oregon, and *Steaming*, published by the National Traction Engine Club of Great Britain.

Much of the growing appeal of steam is based on nostalgia. But there are also enthusiasts who believe there is still a commercial future ahead for steam power in agriculture. Nobody is seriously forecasting that farms in the future will depend upon the massive ploughing and traction engines of the nineteenth century. The great boiler, water tank and coal bunker of the nineteenth-century engine have a future in a museum or at a rally, but not to earn profit on a commercial farm.

Those who hope to farm with steam power in the future are expecting modern technology to adapt the steam-engine to the demands of tomorrow's farming. There is some evidence that this would be possible.

It is quite possible now to design a steam power unit which would eliminate most of the disadvantages of the traditional type of traction engine or portable. One of these disadvantages is the large volume of bulky solid fuel required to operate a traction engine for a day. This posed the problem of supplying coal to the engine, or straw in areas where more efficient fuels were not available, and it also meant that the engine must have a container to carry supplies of the fuel for a journey. Apart from the very rare examples of traction engines equipped with automatic stoking equipment, the engine driver had the job of feeding the fire at intervals during the working day. A steam-engine of the future would use a liquid fuel, and this could be anything which will flow and will burn. A day's supply of fuel would be carried in the tank, and the fuel would be ignited through a burner which would have the fuel supply and flame adjustable according to the work output demanded.

The large boiler of a nineteenth-century steamer, with its bulky size and its high manufacturing cost, would probably be replaced by a tubular boiler. This would contain a very small volume of steam at a high pressure. The result would be greater efficiency and much less weight, bulk and manufacturing cost. An extra bonus would be that the very small volume of steam in a tubular boiler would cause little damage if an explosion occurred. The water in a tubular boiler would be quickly heated, and it would be possible to have full steam pressure in less than 30 seconds from a cold start.

For the engine of a future steam tractor, there would be a choice

164

of designs. The simplest would be a turbine, and this could consist of rotary pistons. The alternative would be a number of reciprocating pistons in a block, and these could be single or double acting.

A new type of steam turbine, publicized in 1975 and developed in Britain, is called OPUS (orbital power unit, steam). This engine has only three main components; these are the pistons and the central shaft to which they are fixed, the rotary valves controlling admission and exhaust of steam and also allowing the pistons to pass, and thirdly the container in which the pistons turn. This design would share the main advantages of turbines' high theoretical efficiency because the pistons travel all the time in the same direction, and also the smoothness which this type of rotary engine would offer. It would be quite possible to operate a steam power unit with water at 1,000 deg. F. and at 1,000 lb per sq. in. pressure, and under these conditions it is claimed that an OPUS unit only 14 in. in diameter could develop more than 800 h.p.

The steam from the engine would be taken to a condenser, and then to a small tank. The tank would act as a reservoir, from which the water would be pumped through to the tubular boiler again for another circuit of the system. The water/steam circuit would be completely sealed, and the amount of water lost in a day's work could be as little as that lost from the radiator of a car. In fact the power unit might not use water, but have instead some other fluid with a low freezing point, to avoid frost damage, and a low boiling point to improve operating efficiency.

A future engine for a steam tractor could have a simple ignition system, probably a battery to operate a glowplug, and this would be needed for starting only. Because of excellent torque characteristics, the steam-engine would probably not need more than a single-forward gear ratio, plus reverse. As the steam-engine will stop and start at will, there would be no need for a clutch.

The reduction in some of the working parts, such as gearbox and clutch, begins to make tomorrow's steam tractor competitive with a modern diesel tractor. There would be other advantages as well, including the extremely quiet running of a steam-engine, absence of vibration if a turbine engine was used, reduced pollution and compact engine size. A steam-engine should have a low maintenance requirement, with fewer working parts. Compared with an internal-combustion engine, where the combustion products are in contact with the piston, cylinder walls and valves, only clean steam would enter the cylinders of a steam-engine, reducing wear and offering many more hours at peak efficiency.

To the steam enthusiast, the advantages of a modern steamer over the conventional internal-combustion engine for tractors and cars, appear overwhelming. Some of these advantages are largely theoretical as comparatively little commercial scale testing has been done recently. But there is also a good deal of fully proven information available. A number of the ideas outlined as possibilities for the future, are simply developments of principles used commercially in steam cars and trucks of fifty years ago.

The motor industry, which produces literally millions of petrol and diesel engines a year for tractors and road vehicles, has invested considerable funds from time to time in evaluating alternative engines and fuels. There appears little doubt that a steam-powered vehicle would be quieter, have a lighter and cheaper engine and transmission, require less maintenance and travel about the same distance on a gallon of fuel, as a conventionally powered vehicle. There have been reports of plans to put steam power into large-scale production, including one report that the Lear corporation of Reno, Nevada, would be producing 1,000 steam-engines a day by 1970, each capable of developing 375 b.h.p. at 10,000 r.p.m., and weighing 65 lb only. So far the potential advantages of steam power for cars and trucks remain unproven on a commercial scale, and reports of large-scale production remain only rumours.

Much of the research has been aimed at putting steam back on to the road, but there have also been attempts to design a steam-powered farm tractor. This research resulted in several prototypes in the early 1920s; at least two companies in the United States demonstrating their steam tractors in public.

The Bryan Harvester Co. of Peru, Indiana, announced a steam tractor in 1922, which was rated at 15 h.p. This tractor, which was quite conventional in external appearance, had a tubular boiler producing steam at up to 600 lb pressure. The overall weight of the Bryan tractor was 5,500 lb, about the same weight as the International Harvester 'Titan' 10–20 h.p., and lighter than the 'Waterloo Boy' 12–25 – two of the well-known petrol–paraffin tractors of the early 1920s. According to Mr R. B. Gray, in his book *Development of the Agricultural Tractor in the United States*, the Bryan steam tractor was the result of a considerable programme of research and development, and the tractor was offered for sale commercially.

The International Harvester Company of Chicago was also experimenting with steam-powered tractors in the period after the First World War. At least two prototypes were built and operated experimentally. Like the Bryan tractor, the International used liquid fuel, a tubular high-pressure boiler with a condenser, and would pull a three-furrow plough. In weight, size, appearance and power output, the Bryan and International steam tractors were completely comparable to most of the tractors available in America at that time. They were also quite unlike the steam traction engines which were still selling in small numbers.

Neither the Bryan nor the International steam tractors achieved commercial success. There is no record of the International version reaching the market, and the company concentrated on its petrol/paraffin tractors, including the Farmall models, which were proving successful against strong competition from the Fordson.

On paper the steam tractor appears an exciting and promising project for the future. In order to translate the paper project into a commercial proposition, the tractor manufacturers will want a little more evidence, more confidence, and a very large budget for retooling and retraining.

Although the tractor manufacturers have been reluctant to offer steam-powered versions of their standard diesel models, it is still possible to buy a brand new traction engine. The manufacturer is the Belmec International Company, based at St Germans, Norfolk. In 1975 the company built two agricultural-type steam-traction engines, in addition to their other engineering work. They hoped that the sale of these two engines might encourage further orders.

Belmec specializes in building completely new traction engines based on the original drawings and patterns used a century or more ago by Savage Bros. Frederic Savage was one of the great pioneers of the traction-engine industry in Britain, and his works near King's Lynn, Norfolk, produced some of the most famous agricultural and showman's engines. The two engines built by Belmec in 1975 were replicas of a chain-driven Savage traction engine built in 1860, and of an 1865 engine.

A full-scale brand-new replica of an historical traction engine is inevitably costly to produce, and the price as this book was written, was estimated at about £17–£20,000 for an engine built to order. Although Belmec had concentrated on Savage engines, they were willing to consider any other make for which original drawings and patterns were available.

Another approach to traction engine building in the 1970s is the do-it-yourself technique of Mr Jerry Crowley of Fenor, Co. Waterford, Ireland. His traction engine, named the 'Leeside Rover' was built by Mr Crowley in 1973. In the same year it showed its paces pulling a three-furrow plough at the World Ploughing Match in Ireland, and was entered at the same event in a competition, sponsored by the *Irish Farmers' Journal*, for new inventions.

The 'Leeside Rover' must be the most unconventional traction engine ever built. The boiler was from a 1936 Marshall, and the single-cylinder engine came from a 1925 Ruston and Hornsby stationary steam plant retired from a creamery. This power unit was mounted on a chassis made up of components from a lorry, including the front axle and rubber-tyred front wheels. The back end of the complete machine came from a Fordson tractor, including the tractor wheels and tyres, the back axle and part of the transmission, and also the original complete three-point linkage. This is probably the only steam-traction engine in the world with a functional hydraulic system and three-point linkage, capable of handling a fully-mounted three-furrow plough.

PLATE DESCRIPTIONS

1 High-pressure engine and boiler built at Hayle, Cornwall, in 1811 by Richard Trevithick. Used for operating a threshing machine and other equipment on a Cornish estate until 1879, when it was presented to the Science Museum in London. Cornish-type boiler delivered steam at about 40 lb per square inch pressure to the 9.5-in. diameter piston. (Illustration based on photographs of the original in the Science Museum.)

2 Colt's Arms Co. engine, patented in America in 1868. Compact design with the engine mounted on top of the vertical boiler. Various sizes available from 2 to 10 h.p.

3 Vertical engine and boiler made by W. Affleck of Swindon, Wiltshire. Rated at $1\frac{1}{2}$ h.p., and advertised in the farming press in 1879, price £55 complete or £20 for the engine only.

4 Vertical engine of oscillating design, and rated at 4 h.p., built about 1865 by William Tuxford and Son, Boston, Lincolnshire.

5 Ferrabee pillar engine, illustrated in *Portable Agricultural Engines* by J. Bourne, 1868.

6 Aimers and Sons of Glasgow stationary engine of *c.* 1878, designed for mounting horizontally or vertically on a wall to save floor space in buildings. Five sizes available, from 4 to 14 h.p.

7 Reading Iron Works horizontal engine, available in the 1870s in a range of sizes from 4 to 16 h.p. The 4-h.p. version cost £55 complete with boiler.

8 Dean portable engine, made in Birmingham in 1845 for the Earl of Craven's farms. A report in the *Farmer's Magazine* in 1845 claimed that a threshing machine driven by the Dean engine had threshed fourteen sacks and two bushels of wheat in 30 minutes.

9 Archamboult 'Forty-Niner', made in Philadelphia in 1849. This was the first agricultural portable steam engine produced commercially in North America.

10   Smith and Ashby $2\frac{1}{2}$-h.p. portable made at Stamford, Lincolnshire, and advertised in 1857 for only £60 complete. A tap on the boiler was designed to allow steam to be piped off to 'cook' animal feeds.

11   Skid-type portable made about 1870 by J. C. Hoadley, Lawrence, Massachusetts. Skid engines were produced in small numbers by several American companies at about this time. They were never a serious competitor for the wheeled type of portable which was easier to move over the long distances between farms when operated by contractors. (Illustration based on photographs of the original in the Henry Ford Museum, Dearborn, Michigan.)

12   Minneapolis portable with return flue boiler, one of a wide range of farm engines made by the Minneapolis Threshing Machine Co. from about 1870.

13   Portable engine with unusual polished metal finish to the boiler, built about 1885 in France by Merlin et Cie., Vierzon.

14   Wolf agricultural engine made at Magdeburg-Buckau, Germany. Built in 1862, this is the oldest German made farm engine surviving, and it was still in regular use in 1904. (Deutsches Museum, Munich copyright photograph.)

15   J. I. Case portable engine built in 1869. This was the first engine built by Case, which later became the world's largest manufacturer of agricultural steam-engines. (Smithsonian Institution, Washington, DC copyright photograph.)

16   Portable engine built by Brown and May, Devizes, Wiltshire in 1880. Single-cylinder engine, 6 n.h.p., weight 96 cwt.

17   Brown and May 3-h.p. portable, pulled by a Shire horse. This engine was built in 1870 and is one of the oldest portables in working order in Britain. Photographed at the 1976 East Anglian Traction Engine Club rally at Weeting, Suffolk.

18   Portable farm engine with vertical boiler, built about 1885 by the Novelty Works of Cory, Pennsylvania. Some American makers persisted with the vertical boiler for several years after almost every British manufacturer had standardized on horizontal boilers for portable and traction engines.

19 Ruston of Lincoln portable built in 1913, with the drawbar converted for hitching to a tractor. (*Farmers' Weekly* copyright photograph.)

20 Marshall portable engine No 70081, built 1916. This engine was sold originally to a timber company in Dublin, Ireland, to operate a saw. (Photograph by Joe Breen.)

21 Portable engine built by Ransomes, Sims and Jefferies, of Ipswich in 1918. Rated at 6 h.p.

22 Ashby, Jeffery and Luke of Stamford, Lincolnshire, portable engine with vertical boiler. An 1876 advertisement for this engine claimed it was 'believed to be the best, safest, most simple and easiest to manage of any in the Trade. They contain few working parts, consume less fuel than any other, give out more power, and cost less money'.

23 Farmer's Engine, 1849, designed and built by E. B. Wilson & Co. of Leeds. Rated at more than 4 n.h.p. and weighing 2 tons 10 cwt, the advanced specification included a sprung rear axle, gear drive with two ratios, and steering from the footplate.

24 Henry Holcroft's optimistic idea for a steam-ploughing engine, designed to propel itself with ploughs attached, by the action of a screw-thread turning against the soil surface. Holcroft described the engine in 1856, and the illustration is based on a drawing published in *The Engineer* of 12 June 1857.

25 Alexander Chaplin & Co. of Glasgow produced a wide range of stationary, portable and self-propelled farm engines about 1861, including this unusual design with a mid-mounted vertical boiler. The traction engine was advertised as available in a range of models from 6 to 27 h.p.

26 Garrett 'Improved Self-Propelling Engine' of 1858. Built at Leiston, Suffolk, this was a chain engine equipped with a traction drive to the rear wheels, and retaining the horse in front for steering only.

27 Atlas self-propelled traction engine of 1881, with steering by means of a horse, and chain drive to the rear wheels. Made by the Atlas Engine Works of Indianapolis, Indiana.

28 The Blackburn ploughing engine of 1857, manufactured in Derbyshire, and demonstrated at the Royal Show at Salisbury, Wiltshire. Designed for ploughing, with the 8-ft diameter drum intended to give good traction with minimum soil compaction, the machine was offered for sale for £460.

29 Several patents were taken out for 'portable railway' track systems, following the temporary sucess of the Boydell system. This is the Remington system, patented in England in 1857.

30 Stratton traction engine, built 1893 in Moscow, Pennsylvania. The half-track design was intended to provide traction on cultivated soil, and there was a shaft from the engine to act as a power take-off for stationary work.

31 D. June & Co. traction engine of 1899, made at Fremont, Ohio. The pot-shaped container on top of the boiler holds water, and was a patented system for preventing sparks leaving the chimney to create a fire hazard when threshing.

32 Collinson Hall and Thomas Charlton ploughing engine of 1857. Advanced features included a patented safety device for steep ground, to ensure that steam and not water went into the cylinder; provision for two-way working with the steering mechanism having a steering wheel at each end; and a steam-powered ram which pushed down beneath the boiler to act as a jack to enable the machine to be turned manually at the headland.

33 J. I. Case Threshing Machine Co. of Wisconsin straw-burning traction engine of 1885. This model was self-propelled and equipped with steering operated by the steering wheel at the footplate. There was also provision for horses to pull and steer the engine, and this accounts for the seat and footrest at the front.

34 Steam-powered rotary cultivator patented in 1860 in America and described as the 'Gatling Machine for Pulverizing the Soil'. Oxen or horses were used to pull the machine, avoiding the problems of self-propulsion on cultivated land.

35 Platt rotary plough, patented 1858 by Henry Platt of New York. Small versions were to be pulled by horses, and larger models by steam. As the plough wheels turned they operated the rotary plough body in a corkscrew motion intended to plough and cultivate in one pass.

36 Hop-digging machine patented by John Knight of Farnham, Surrey, and described in the 1 July 1876 edition of the *Implement Manufacturers' Review*. The digger was powered by a portable engine at the headland, through a cable system.

37 Rotary plough patented 1849 by James Usher of Edinburgh. The illustration is of a contemporary model of the steam plough, which is now in the Science Museum, London. The full-scale Usher plough weighed $6\frac{1}{2}$ tons and was powered by a 10-h.p. engine. (Crown copyright photograph.)

38 Thomas Rickett of Buckingham self-propelled rotary cultivator, entered for the 1858 trials of the Royal Agricultural Society. The cultivating width was 7 feet, and the outfit was claimed to cover 4 acres a day tilled to a depth of 6 inches.

39 Monckton and Clark rotary cultivator patented in London, 1856. The rotary cultivator simply tilled a strip of ground beside the machine as it moved forward.

40 Monckton and Clark spading machine, described 1857. A series of levers actuated the spades or digging bodies within the wheelbase of the machine as it propelled itself slowly forward.

41 Grimmer ploughing engine, built at Walsoken, Lincs in 1881. This was an attempt to build a general-purpose power unit for haulage, cultivating and stationary work. Rated at 4 n.h.p. and weighing only $2\frac{1}{2}$ tons, the aim was to make the machine cheap enough for medium-sized farms to buy. The rear-wheel steering was simple and probably effective, but the drive mechanism to the two front wheels probably proved troublesome.

42 Frank Proctor's digging attachment for various makes of traction engine. This was commercially the most successful of the many nineteenth-century attempts to build a steam-powered rotary cultivator or spading machine. Many of the Proctor units were exported during the 1880s.

43 Savory ploughing engine built by W. Savory and Son of Gloucester and demonstrated successfully in 1863. The winding drum for the plough cable was 6 ft diameter, and revolved around the boiler.

44 Burrell and Sons of Thetford, Norfolk, 'Universal'-type plough-ing engine advertised in the early 1880s. The vertical winding drums were mounted on each side of the boiler of a basic 10-h.p. Burrell traction engine. The 'Universal' engines were designed to operate with various makes of single-engine cable-ploughing tackle.

45 Fowler double-engine ploughing system demonstrated at the 1976 East Anglian Traction Engine Club rally, with a pair of 1921 Fowler BB1 engines, Nos 15420 and 15421, with a Fowler six-furrow balance plough.

46 Fowler BB1 ploughing engine, 1921, owned by Mr K. Steward of Cockfield, Suffolk.

47 Two-way cultivator by Howard of Bedford, produced in the 1860s and 1870s for use with cable-ploughing equipment. The implement is designed for two-way working, with a seat and steering control at each end.

48 Cultivator designed in 1862 for steam cable operation, by William Smith of Woolston, Buckinghamshire.

49 Plough for heavy duty reclamation work on moorland, built by Fowler of Leeds for the estates of the Duke of Sutherland in Scot-land. The roller wheels were designed to stop the plough sinking in soft soil. In limited production from 1871.

50 Portable anchor for steam cable-ploughing systems, designed by F. Savage of King's Lynn, Norfolk, in 1879.

51 Knights and Stacey mole plough, sold in the 1880s for use with steam-cable systems, and said to be suitable for operation with a 12-h.p. engine.

52 Windlass to operate a single-engine cable-ploughing system, designed in 1855 by John Williams of Baydon, Wiltshire. The wind-lass was intended to be used with a portable engine, and for opera-tion with a belt drive.

53 Subsoiling plough for use with a steam cable system, designed about 1865 by William Smith. The tine behind the plough body acted as a pan-buster to break up compaction in the furrow bottom.

54 Water-tender produced towards the end of the nineteenth century by the J. I. Case company in America.

55 S. T. Osmond of Ramsbury, Wiltshire, was one of many country engineering firms to produce special equipment to take advantage of the steam farming boom in the nineteenth century. Osmond produced water carts as a speciality, with a wide range of sizes and designs. In 1879 their two-wheel, 120 gal. tank, cost £13, and the largest model, with four wheels and 200 gal. capacity cost £23.

56 Early type of stationary baler or straw press made from 1881 by J. & F. Howard of Bedford, designed to work with a portable engine.

57 Ann Arbor Machine Co. of Ann Arbor, Michigan, claimed that their 'Columbia' model stationary baler held the American record of 68 tons of hay baled in 10 hours in 1904.

58 Stationary baler for making round, or cylindrical, shape bales designed by M. Th. Pilter of Paris in the late 1870s. The baler was operated from a steam-engine, and the dense bales produced were said to be ideal for long-distance transport. The French cavalry horses in the 1880s were said to have been supplied with hay from a Pilter baler.

59 The 'Perpetual' baler, patented in America, but imported into Britain in 1881. Operated with power from a steam-engine, the machine could produce bales of varying length, high density, and up to almost 2 cwt.

60 Aveling and Porter traction engine with a Crosskill reaper, demonstrated in Warwickshire in 1876.

61 Best combine harvester towed behind a Best traction engine. These harvesting outfits were used in the 1890s for the extensive grain acreages of California.

62 Case 'Agitator' threshing machine of 1897, designed for very high output, and complete with a wind-powered stacker for the straw.

63 Another version of the Case 'Agitator', with an elevator-type straw stacker.

64 The teams of men who travelled with steam-ploughing rigs on contract work used living vans, such as this one manufactured by Fowler of Leeds. The vans were complete with cooking stoves, beds, and stowage for tools and spare plough points.

65 A 'big-wheel' steam-traction engine manufactured in 1900 by Best for a farm in California. The extra-wide wheels were to prevent the 41-ton monster sinking in soft ground. Each of the driving wheels was 9 ft high and 15 ft wide.

66 Replica of a 1865 traction engine by F. Savage of King's Lynn, Norfolk. This modern version, built in 1975 by Belmec International Ltd, is based on the original drawings and patterns and represents one of the earliest commercial self-propelled engines.

67 Marshall of Gainsborough, Lincolnshire, 7 n.h.p. traction engine of 1920, working with a Powell stationary baler, in a threshing demonstration at the 1976 rally of the East Anglia Traction Engine Club at Weeting, Suffolk.

68 Fisher-Humphreys stationary baler, designed to make high-density, wire-tied straw bales. This machine, dating from the 1930s, has been fitted with more modern road wheels.

69 Ransomes, Sims and Jefferies threshing drum, built about 1938, photographed at the 1976 Banbury, Oxfordshire, Steam Rally.

70 Tree felling saw, powered by steam and announced in 1878 by A. Ransome and Co. of Chelsea, London. At a public demonstration the saw cut through the 3-ft diameter trunk of an oak in 4 minutes.

71 Buckeye Traction Ditcher Company of Findlay, Ohio, advertised this steam-powered trench digger for tile drainage in 1910. It was claimed to dig up to 150 rods a day to a depth of 2/3 ft, ready for tiles to be laid.

72 Robson and Herdman draining machine, 1881, built at Newark, Nottinghamshire. Pulled by a cable from a steam-engine drum, two bucket conveyors dig a trench, pipes are laid in the trench, and the soil is then returned to cover the pipes.

73   Chicago Farm Tile Ditcher, manufactured in the early 1900s by the Municipal Engineering and Contracting Co. of Chicago, Illinois. The wide wheels at the front and tracks at the rear helped the machine over wet land. A gasoline or petrol engine was available as an option by 1908.

74   Westinghouse traction engine, with vertical boiler, manufactured in Schenectady, New York. This basic design, with some modifications and improvements, was in production for more than thirty years from about 1881.

75   Burrell 8 n.h.p. traction engine, made at Thetford, Norfolk, in 1876, and thought to be the oldest Burrell in working order.

76   Shaft-drive traction engine produced by the Aultman-Taylor Co. of Mansfield, Ohio, in the 1880s. The massive bevel gear behind the flywheel was known as a 'daisy gear'. The shaft drive was moderately popular, but the more conventional gear drive proved more reliable.

77   Marshall of Gainsborough traction engine, built 1887, No 15391, rated at 6 n.h.p.

78   Avery 40 h.p. undermounted traction engine, produced in the early 1900s. Working parts of the engine were easily accessible for maintenance, but the design lacked stability on steep ground.

79   'Rubicon' four-wheel drive traction engine produced about 1885 by Wood, Taber and Morse of Eaton, New York. This was an attempt to achieve more efficient traction, but the extra complexity of the transmission and steering were a disadvantage. (Illustration based on photographs of the engine at the Henry Ford Museum, Dearborn, Michigan.)

80   Gaar Scott 'Big Forty' traction engine, rated at 120 b.h.p. and 40 h.p. at the drawbar. Weight with full water tanks was 35,000 lb. Built at Richmond, Indiana, from about 1905.

81   J. I. Case 28–80 traction engine of 1912, delivering 28 h.p. at the drawbar, and suitable for direct ploughing as well as stationary work.

82   Marshall traction engine, built in 1908, and at one time owned by the Oxford Steam Ploughing Co. of Middle Cowley, Oxford, one of the leading contractors in the steam ploughing business.

83 Burrell 6 n.h.p. traction engine, built at Thetford, Norfolk in 1901. Engine No 2426, weight 11$\frac{1}{2}$ tons.

84 Wallis and Steevens of Basingstoke, Hampshire, traction engine No 7685, single-cylinder 6 n.h.p., built in 1919.

85 Garrett 7 n.h.p. traction engine built at the Leiston, Suffolk, factory in 1916.

86 Burrell 'Devonshire' engine, single-crank compound design of 1909, rated at 6 n.h.p. This model was specially developed as a lightweight traction engine with ample power for hauling heavy machinery in hilly areas such as Devonshire.

87 Allchin 7 n.h.p. general-purpose traction engine, built at Northampton in 1911. The engine number is 1546, and it is a single-cylinder design.

88 Clayton and Shuttleworth 6 n.h.p. traction engine built in 1923, and representative of the conventional type of general-purpose engine built towards the end of the steam farming period.

89 Ruston Proctor 7 n.h.p. agricultural traction engine of single-cylinder design, built in 1908 with the works number 34987.

90 Burrell of Thetford, Norfolk, 8 n.h.p. single-cylinder agricultural traction engine, built in 1907.

91 Burrell traction engine No 3923, built 1922. This is a 7 n.h.p. double-crank compound design, built for the Burrell stand at the 1922 Royal Show at Cambridge.

92 Fowler of Leeds 3$\frac{1}{2}$ n.h.p. traction engine built in 1921.

93 Marshall of Gainsborough, Lincolnshire, 6 n.h.p. general-purpose traction engine No 71837. Built in 1919, this engine weighs 10 tons.

94 Ruston Proctor Agricultural engine No 50278, built 1914. This engine has been owned, mainly for contract work, by generations of the same family since new. In 1914, when requisitioned for war work, this engine established what was then a national record by baling 242 tons of hay through a stationary baler in one week.

95 Ransomes, Sims and Jefferies of Ipswich 7 n.h.p. traction engine of 1919.

96 A parade of engines at the East Anglian Traction Engine Club 1976 rally at Weeting, Suffolk.

97 and **98** Two prototype steam tractors produced in America in 1922 and 1923 by the International Harvester Co. of Chicago, Illinois, Both tractors were operated experimentally, and were used for ploughing and other field work, but neither was developed commercially.

99 Bryan steam tractor, made at Peru, Indiana, in 1923. This was an attempt to build a farm tractor to compete with the Fordson and International Harvester models, but with a high-pressure tubular boiler and kerosene burner as the power unit. (Illustration by John Wood, taken from *Farm Tractors in Colour* by Michael Williams.)

100 Sentinel Roadless steam tractor, 1924. This was a technically advanced tractor, built at Shrewsbury, Shropshire, by a company better known for steam waggons. The tractor was developed for land reclamation work and cultivations in East Africa, and was claimed to have a 6-ton pull at the drawbar.

101 American Abell traction engine of 1911, offering 32 h.p. at the drawbar for ploughing, and 120 h.p. at the belt. The total weight was 25 tons; the tricycle wheel arrangement was never popular.

102 The Garrett 'Suffolk Punch' offering such advanced features as super-heating and Ackerman steering in 1919 – too late to fight off the challenge of the farm tractor.

103 Jerry Crowley's 'Leeside Rover' traction engine, built at Fenor, Co. Waterford, Ireland in 1971, and including parts of a 1925 Ruston and Hornsby engine, a 1936 Marshall boiler, front wheels from a lorry and the rear end of a Fordson tractor. This is probably the only steam-traction engine with hydraulically operated three-point linkage. (Photograph by Joe Breen).

INDEX

(Plate references are set in bold type.)